Heart of Science

Heart of Science

A Philosophy of Scientific Inquiry

JACOB STEGENGA

THE UNIVERSITY OF CHICAGO PRESS   CHICAGO AND LONDON

The University of Chicago Press, Chicago 60637
The University of Chicago Press, Ltd., London
© 2026 by The University of Chicago
All rights reserved. No part of this book may be used or reproduced in any manner whatsoever without written permission, except in the case of brief quotations in critical articles and reviews. For more information, contact the University of Chicago Press, 1427 E. 60th St., Chicago, IL 60637.
Published 2026
Printed in the United States of America

35 34 33 32 31 30 29 28 27 26   1 2 3 4 5

ISBN-13: 978-0-226-84403-9 (cloth)
ISBN-13: 978-0-226-84405-3 (paper)
ISBN-13: 978-0-226-84404-6 (ebook)
DOI: https://doi.org/10.7208/chicago/9780226844046.001.0001

Library of Congress Cataloging-in-Publication Data

Names: Stegenga, Jacob, author. https://id.oclc.org/worldcat/entity
　/E39PCjxPj33MJMyHGtWGmgrMKd
Title: Heart of science : a philosophy of scientific inquiry / Jacob Stegenga.
Description: Chicago : The University of Chicago Press, 2026. |
　Includes bibliographical references and index.
Identifiers: LCCN 2025027372 | ISBN 9780226844039 (cloth) |
　ISBN 9780226844053 (paperback) | ISBN 9780226844046 (ebook)
Subjects: LCSH: Science—Philosophy.
Classification: LCC Q175 .S756 2026 | DDC 501—dc23/eng/20250613
LC record available at https://lccn.loc.gov/2025027372

♾ This paper meets the requirements of ANSI/NISO Z39.48-1992 (Permanence of Paper).

Authorized Representative for EU General Product Safety Regulation (GPSR) queries:
Easy Access System Europe—Mustamäe tee 50, 10621 Tallinn, Estonia, gpsr.requests @easproject.com

Any other queries: https://press.uchicago.edu/press/contact.html

ANASTASIIA—HEART OF BRIGHTNESS—CREATED THE CALM AND CARING CONDITIONS FOR ME TO WRITE. THIS BOOK IS DEDICATED TO HER.

Contents

INTRODUCTION   1

CHAPTER 1. Common Knowledge   26

CHAPTER 2. Deontic Evaluation   62

CHAPTER 3. A New Value-Free Ideal   87

CHAPTER 4. Scientific Assertion   116

CHAPTER 5. Scientific Progress   146

CHAPTER 6. Prise Praise and Prize from Priority   166

CHAPTER 7. Fast Science   183

CHAPTER 8. Timeless Truths   208

CONCLUSION   229

Acknowledgments   231

References   233

Index   251

Introduction

Heart of Science

The heart of science is justification. Justification provides the oxygen for all other aspects of science. This book offers a philosophical view of science based on the centrality of justification.

The ultimate aim of science goes far beyond justification. The aim of science is a special kind of truth, which I call *common knowledge*, a shared and mutually justified scientific finding. Justificatory practices contribute to the achievement of common knowledge, but evaluative concepts for assessing various aspects of science—scientific progress, creditworthy science, or appropriate scientific testimony—should not be based on whether the scientific work under evaluation has achieved its aim. Rather, we should base those evaluative concepts on the extent to which the scientific work under evaluation has justified its claims. Our concepts for judging science should be process-oriented rather than product-oriented. Our evaluation of science should, to use philosophical terms of art, be *deontic* rather than *consequentialist*. Good science need not attain its aims, it must justify its claims.

Consequentialist, product-oriented philosophy of science has focused on the ability of science to achieve its aims, which can include factive aims such as truth or knowledge, or other kinds of successes such as problem-solving capacity. Take the notion of scientific progress, for example. A product-oriented account of scientific progress holds that science makes progress when it accumulates truth or knowledge, whereas a process-oriented account holds that science makes progress when it properly deploys its justificatory practices. While properly deploying justificatory practices can lead to truth, our judgment that science has made progress need not depend on that success. Physicist Lise Meitner, who contributed to the discovery of nuclear fission, claimed that science is "a battle for final truth" (Rhodes 1986, 234). More ironically, Friedrich Nietzsche (1968, 252) wrote of "the lovely paths of truth." While science can very often win the battle for final truth or get to the end of a particular lovely path of truth, our evaluative concepts for science should be focused on the battle rather than the victory, on the lovely paths rather than the ramble's end.

An evaluative concept is one with normative content. In the context of science, an evaluative concept can be used to judge the quality of some aspect of science; examples of evaluative concepts for science include scientific progress, scientific credit, and norms of assertion for science. Placing justification at the center of our evaluative concepts for science allows the formulation of standards of judgment that can be met by actual science in real time as science plays out. Standards of judgment that are too demanding can be met only by an idealized science, or science of the future. Yet standards of judgment based on justification are not so easy to meet such that nonscientific enterprises can easily satisfy them precisely because justificatory practices in science are demanding.

What is scientific progress? What features should good scientific testimony have? What is creditworthy science? Can we demarcate good science from bad? Should scientific reasoning be purified of value-laden influences? Should science sacrifice its standards to quickly respond to a great threat such as a pandemic? In this book, I offer novel answers to these questions by emphasizing the centrality of justification and justificatory practices—the heart of science.

There are two general ways in which my focus on justification and justificatory practices forms the basis of novel answers to a range of questions in philosophy of science. First, the centrality of justification displaces truth or knowledge as necessary components of evaluative concepts. Truth and knowledge are the basis of "factive" accounts of evaluative concepts. For example, a factive account of appropriate scientific testimony holds that scientific testimony is appropriate only if that which is asserted in scientific testimony is known or true. I resist factive accounts of norms of scientific testimony, scientific progress, and scientific credit, and in their place, I offer justification-centered accounts. The problem with factive evaluative concepts for science is that, to put it bluntly, the point of doing science in the first place is to discover truths, yet the determination that a scientific conclusion is true or, conversely, that a hypothesis or guiding research framework is false can take many years or even decades or centuries, and thus in real time, truth cannot be the basis of prescriptive action-guiding norms or evaluative norms. Yet, again, to put it bluntly, the point of evaluative concepts and norms is to guide action and evaluation. It follows that evaluative concepts and norms for science should not be factive.

Second, the deontic philosophy of science pursued here, with its focus on justification and justificatory practices, offers new ways of characterizing important notions in philosophy of science and fresh arguments and novel positions on familiar debates, including values in science, the aim of science, the demarcation of science from pseudoscience, and the exploration of topics that are, for philosophers of science at least, quite new, such as science in a time of crisis. For example, philosophers have nearly reached a consensus after decades of scholarship—reignited by an important article by Douglas (2000)—that nonscientific values can and should influence the internal workings of scientific reasoning. Dissenters resist this troubling conclusion, but they often neglect compelling arguments for it. Yet the entire literature—both by defenders of the status quo and by dissenters—has been consequentialist, arguing whether values do or must or should permeate the end-state of scientific conclusions. A deontic, non-consequentialist approach to this debate provides a fresh outlook. This approach grants that the end-state of scientific reasoning is or must be or should be value-permeated, based on the well-known argument from inductive risk, while maintaining that scientists should strive to eliminate value-permeation by deploying and developing their justificatory practices as far as possible. This argument is developed in chapter 3, which I wrote with my friend Tarun Menon during three years of collaboration.

Though I argue that neither truth nor knowledge should be the basis of evaluative concepts for science, that does not imply that truth and knowledge are unimportant for science. Far from it. I agree with Bird (2022) that science aims at knowledge and is often successful at attaining knowledge, though in chapter 1, I supplement this view by arguing that the aim of science is a special kind of knowledge, namely, common knowledge. Since knowledge entails truth, I concur that an aim of science is truth, a special kind of truth that is justified by rigorous scientific practice, and those justifications are scrutinized, improved, and ultimately accepted by the broader scientific community. Yet from this, it does not follow that the achievement of truth in any particular instance of scientific work should be the standard that we build into our evaluative concepts for science.

That would be so only on a consequentialist philosophy of science. A deontic philosophy of science uses justification and justificatory practices rather than truth or knowledge as the standard for articulating evaluative conceptions for science.

The two most famous philosophers of science and certainly the two who have had the most impact on the way scientists think about their work and the way society thinks about science are Thomas Kuhn and Karl Popper. Many readers interpreted the central message of Kuhn's (1962) *The Structure of Scientific Revolutions* to be that scientific disciplines routinely go through major upheavals that involve discarding existing theories and replacing them with completely new theories. The central message of Popper's (1963) *Conjectures and Refutations* was that scientific theories never receive any confirmation, should never be deemed to be true, and should always be entertained as provisional hypotheses while scientists do all they can to falsify them. These two messages have fermented into the toxic brew that I call the provisionality thesis, which says that all scientific theories are provisional, and thus science can never claim to have achieved truth.

In academia, that toxic brew was popular in the nineties and contributed to the so-called science wars. Most responsible scholars have worked off the hazy hangover from this brew (though many academics stubbornly promulgate it). Yet a similar brew seems to have poisoned society. If science does not attain truth and everything is a matter of perspective, then we are in a post-truth era of "alternative facts." School boards put creation science beside evolutionary theory on curricula, corporations publish only convenient data about the drugs they sell, large numbers of people doubt science's best accomplishments, and politicians enact policies based on absurdly shoddy science yet take little action on issues based on extremely compelling science. Scholars respond by encouraging the opposite extreme, urging people to trust science (Oreskes 2019); even scholars who once told us to doubt science now encourage trust in science (Latour 2004; Collins and Evans 2017). Trust, though, must be earned. Notice what results from believing both the provisionality thesis and a consequentialist philosophy of science: It is hard to see why we should trust science if the achievement of truth is our basis for judging science and science cannot achieve truth. Both the provisionality thesis and a consequentialist philosophy of science should be rejected. Science can achieve truth and knowledge, yet those achievements should not be the basis of trust in science or any other mode of judging science. Trust in science, like other evaluative concepts for science, should be process-oriented, based on the quality of the justificatory practices of science.

I wrote half of this book while living in Gottfried Leibniz's former house in Hanover (a reconstruction of his house, as the original was destroyed by the Allied bombings during the Second World War). A young Leibniz wrote that "the sole end of philosophising" was "for use in life and to increase the power and happiness of mankind" (quoted in Antognazza 2008, 88). Channeling the spirit of Leibniz, I believe that a philosophical conception of science is only worth pursuing if it can be relevant to science or the use of science by society, perhaps to increase the power and happiness of humanity, as long as those ends are properly construed. An enduring tradition in philosophy holds that philosophical ideas should improve our lives and the world around us. This book is part of that tradition.

I wrote the other half of this book while living in Cambridge, working out of my office just a few meters from the former site of the Cavendish Laboratory, where J. J. Thomson discovered the electron, James Chadwick discovered the neutron, and James Watson and Francis Crick discovered the structure of DNA. Having had the privilege of working in the same institution as Isaac Newton, Jane Goodall, and Stephen Hawking, I held myself to a working assumption that science at its best should be able to satisfy the evaluative concepts we use to judge it. A consequentialist philosophy of science promulgates evaluative standards that not even the very best science can satisfy, while a deontic philosophy of science articulates standards that good science can and regularly does meet. Assessing the health of science should be based on evaluating the heart of science.

Deontic Philosophy of Science

Suppose you are the parent of a child who very much wants to be a famous violinist. As a loving, supportive parent, you nurture the pursuit of this goal, despite knowing your child will likely not achieve it. You pay for the best instructors and buy your child the best violin that your budget allows. Your child practices in every spare hour, for years—practicing scales, memorizing complex sequences, developing manual dexterity. Yet after all this effort, your child does not achieve her goal. She becomes a fine musician, learns much about music, and impresses friends at dinner parties, but she does not quite make it to the Royal Albert Hall. Your child, of course, is disappointed. As a loving parent and mature observer, *you* are not disappointed. You are proud. You witnessed your child display great effort and devotion and develop a wonderful talent that has enriched their life and character. I hope you have the intuition that there is absolutely no sense in which your child has failed.

Consequentialism about musical aims would say that since your child has not achieved their goal, they have failed. That is a serious mark against consequentialism about musical aims. Nonconsequentialism, however, agrees with the intuition that your child has not failed. Our judgment should focus on the procedural steps one takes to pursue a goal—judging whether they are good steps and whether they are performed well—and not on whether the goal was attained.

Now consider an alternative story about your child's musical development. Suppose that early in her teenage years, your child reads about a new technology that allows a person to absorb the talents of another person by giving that person a special drug and then connecting the two of them through a dialysis machine. Your child secretly devises a plan. She kidnaps a famous violinist, forces the violinist to take this drug, and then connects to them via a dialysis machine, against the violinist's will. Your child absorbs all of the violinist's musical talent, which allows her to achieve her goal. She no longer practices much, but she still becomes a famous violinist. You are, of course, proud. But you later learn about her scheme. Your pride—I hope and assume—turns to dismay, despite the fact that your child has achieved their goal.

Again, consequentialism about goal-seeking gives the wrong verdict, while nonconsequentialism about goal-seeking gives the right verdict. Using Bradford's (2015) insightful analysis of the notion of achievement, the child's outcome in the second scenario is not an achievement because, by cheating, the path to her goal is no longer difficult; indeed, it is trivial.

These two stories should motivate our deontic intuitions. A narrow construal of the term *deontic* implies a concern with duties and obligations, while a broader construal implies, in addition, a concern with permission and recommendation, and involves the related notions of right and wrong, ought and can, evaluative responses of praise and blame, and judgments of better and worse. Deontic evaluation is about judging the rightness of an action by reference to rules of obligation or permission and conditional claims about the best means to reach a goal—claims of ought, can, should, or must. Some quotidian examples of deontic evaluation include: "Dogs are allowed in this café," "Russia should not have invaded Ukraine," "If you want good sushi, you should go to that restaurant," and "She studied diligently for the logic exam." While consequentialist evaluation judges the goods and harms that result from action, deontic evaluation judges the goodness of action itself, asking if an action accords with principles of obligation, permission, or prudence.

Deontic philosophy of science maintains that our evaluative concepts for science should not be based on the achievement of the goal of inquiry, namely, truth or knowledge. Philosophers are familiar with such a deontic notion, as it is borrowed from ethics. Consequentialist ethics holds that appropriate action in any given situation is that which will bring about the best consequences, while, in contrast, deontological ethics holds that appropriate action is that which best accords with principles of duty or obligation. Just as deontological ethics holds that evaluation of action should be based on the extent to which an action follows apt principles or norms rather than the extent to which an action leads to positive consequences, deontic philosophy of science holds that evaluation of science should be based on the extent to which particular scientific work follows apt principles or norms rather than whether that scientific work led to truth. Of course, principled actions very often lead to positive consequences, and principled scientific work very often leads to truth. Yet our evaluative concepts should be based on accordance with principles or norms rather than attainment of truth.

Deontological ethical principles are familiar: Do not lie, do not steal, treat people with respect. These principles differ slightly among cultures. In the traditional West African Akan culture, ethical principles include hard work, hospitality, and respect for elders. In Christian traditions, ethical codes are presented in the Bible in both narrative manner and enshrined codes such as the Ten Commandments (don't kill, don't steal, honor your parents . . .). In the Haida culture indigenous to the west coast of Canada, ethical principles include showing respect for all living things, the importance of advice from elders, and reciprocal gift-giving.

What are the analogous principles for a deontic philosophy of science? These are any principles or practices that contribute to justifying a scientific claim. Justificatory principles and practices are those that increase the reliability or objectivity of science (though, as I argue in chapter 1, one may want more than mere truth-conducive reliability for justificatory principles and practices). These include domain-general principles, applying to all of science, and domain-specific practices, applying to one part of one discipline. Consider first some candidate domain-general principles. These include any norm of reasoning more generally, including logical rules such as modus ponens and norms of probabilistic inference such as the norm that one should not commit the base-rate fallacy. There are also domain-general norms that are specific to science. Sociologist Robert Merton ([1942] 1973) argued that science is governed by a set of norms that include universalism, organized skepticism, communism, and disinterestedness. Famously, Popper argued that a scientific theory should be falsifiable. More recently, Longino (1990) argued that structured practices of criticism at the community level allow science to attain objectivity. Consider now some candidate domain-specific principles. To test the effectiveness of a new pharmaceutical, one should perform a randomized controlled trial (this norm has been codified as a strict obligation by regulatory agencies). To predict the spread of a pandemic, one should use an epidemiological model, get as much empirical evidence as possible to set the parameters of the model, and perform sensitivity analyses by running multiple simulations while varying the values of the parameters one is uncertain about. Such local principles can of course be technical and boutique: For example, to correct for the epistemic risk of performing multiple data analyses, one should perform a so-called Bonferroni correction. Throughout this book, I refer to domain-general principles as *justificatory principles* and domain-specific practices as *justificatory practices*, and I use the term *justificatory norms* to refer to both. Justificatory norms are the heart of science.

To be clear, I do not mean norm in the statistical or conventional sense but rather in a fully normative sense—for example, while peer review is a very common practice in science, there is some debate about whether that practice is, on the whole, a good thing for science and thus whether there is a genuinely justificatory norm that scientists submit their work to peer review (Heesen and Bright 2021). In chapter 1, I describe the fundamental basis of the normative status of justificatory norms, and in chapter 2, I use the notion of justificatory norms to offer a new approach to scientific demarcation.

In the previous section, I noted several examples of topics that can be illuminated by a deontic philosophy of science, including the so-called value-free ideal (chapter 3) and the notion of scientific progress (chapter 5). Here is another, which I discuss in chapter 4: Under what conditions are scientific assertions appropriate? That is, if a scientist makes a claim about the world, such as "atoms have a dense nuclear core" or "masks mitigate the spread of a respiratory virus" or "the universe is expanding," under what conditions should we respond with approbation and under what conditions should we respond with condemnation? Using a term of art from epistemology, this question asks: What are the "norms of assertion" for science? Surprisingly, despite the supreme importance of scientific assertion for society, philosophers of science have had little to say about this question, while epistemologists have offered convoluted or implausible views about the norms of assertoric discourse in general. Many epistemologists tell us that an assertion is appropriate only if it is true or known (the most prominent defense of a knowledge norm of assertion is offered in Williamson [2000]). A deontic philosophy of science tells us to focus on the quality of the means (justification) rather than the attainment of the end (truth or knowledge). The norm of assertion, according to a justification-centered approach (both for science and assertoric discourse in general), is that to be appropriate, an assertion must be justified—an asserter must have good reasons that justify their assertion. A justification norm of assertion has crucial advantages over factive norms (truth or knowledge): The justificatory status of an assertion is epistemically accessible at the moment of assertion, while the truth of an assertion is, in most scientific contexts at least, ascertainable only in retrospect; a justification norm of assertion gives clear guidance to scientists, while a factive norm does not; and a justification norm closely tracks scientific practice itself.

The analogy between deontological ethics and deontic philosophy of science might raise concerns among readers familiar with standard objections to deontological ethics. Let us consider one here. One of the best-developed deontological systems of ethics is due to eighteenth-century philosopher Immanuel Kant. A feature of Kant's ethics is that it is never permissible to lie to anyone for any reason. Critics heaped scorn on this. Contemporaries of Kant posed a now-famous thought experiment: Suppose a murderer is looking for a friend of yours who is hiding in your house, and the murderer knocks on the door and asks you if your friend is there. What should you do? The intuition the thought experiment attempts to elicit, of course, is that it is not just permissible to lie to the murderer, it is your obligation to lie. That intuition might be motivated by the thought that lying will bring about the best possible outcome in this difficult situation, and if so, your intuition fundamentally favors consequentialist ethics rather than deontological ethics.

However, the analogical scenario in science is not a challenge to a deontic philosophy of science. When faced with the murderer at the door, the thought experiment assumes that you can determine which action will lead to the best consequences or at least which action could be expected to lead to the best consequences. The analogical scenario in science would involve a scientist who can violate a justificatory norm and in doing so would discover a truth that she would not have otherwise discovered—or at least increase the chance of such a discovery. Surely, there are many real examples in the history of science of scientists breaking conventional rules and thereby discovering truths (this is fundamentally the complaint made by Lakatos and Feyerabend against Popper). Yet, in such scenarios, a scientist cannot determine that violating a justificatory norm would lead to truth; indeed, they have every reason to suppose otherwise since justificatory norms are generally truth-conducive. Of course, a scientist might believe that a putative justificatory norm is not, in fact, truth-conducive, in which case she might believe that violating the norm could lead to truth, but this would amount to the scientist calling into doubt the status of the justificatory norm itself. Moreover, it is doubtful whether a particular scientist can reliably predict that by violating a specific justificatory norm, they will be more likely to attain truth (which is thus very much unlike analogous cases in ethics, such as the murderer at the door).

One might say that the analogy between deontological ethics and deontic philosophy of science is strained because in deontological ethics, norms such as "don't lie" get their status as principles because they are derived from more fundamental principles, such as the principle that all people must be treated as ends in themselves rather than mere means, and not because such principles necessarily contribute to positive outcomes. In deontic philosophy of science, however, justificatory norms get their status as principles not because they are derived from some more fundamental principle but because they are truth-conducive, goes this thought. The fundamental epistemic good is truth (or knowledge), and justificatory norms have normative status only insofar as satisfying those norms is a means to the end of attaining truth. Thus, one might say, deontic philosophy of science amounts to a version of "rule consequentialism" since it is ultimately the consequence of truth-attainment that grounds the normative status of justificatory norms. In ethical theory, rule consequentialism is the view that we should follow a specified set of rules because doing so tends to lead to the best consequences. The first strain here is that if the deontic principles are not derived from a more fundamental principle but rather get their status as principles only on the grounds that they are truth-conducive, then deontic philosophy of science amounts to rule consequentialism and so would be, after all, consequentialist and not deontic. The second strain is that in ethical theory, rule consequentialism might collapse into act consequentialism because, for any given action, if one could break a rule to achieve better consequences, then one ought to break the rule, and to uphold the rule for this action would amount to "rule worship" (Arneson 2005). If so, then deontic philosophy of science might also collapse into act consequentialism with respect to the aim of truth.

However, a crucial difference between the ethical domain and the scientific domain is that in the ethical domain, as we saw with the murderer at the door, we can sometimes predict that rule violations will increase the chance of better consequences, but in the scientific domain, it is much harder to reliably predict that violating a justificatory norm will increase the chance of discovering a truth because there is a tight relationship between the normative status of justificatory norms and truth-conduciveness (as I will argue in chapter 1). So, scientific rule consequentialism is not faced with a collapse into scientific act consequentialism. Nevertheless, one might think that the normative status of justificatory norms is fundamentally due to their truth-conduciveness, and so one might insist on calling the view here a rule consequentialist philosophy of science rather than a deontic philosophy of science.

Yet that would ignore what grounds the value of truth and knowledge itself. As Chrisman (2022, 4) argues, "Knowledge is valuable in large part because of the stabilizing role it plays in social cooperation." Pragmatists have long argued that truth *simpliciter* is unimportant—the number of grains of sand on Tamarind Beach in Bali is not worth knowing. Only "significant" truths are important, a claim compellingly argued by Kitcher (2001). This is not to downplay the genuinely epistemic achievement of knowledge; on the contrary, it is the importance of that epistemic achievement that contributes to its stabilizing role in social cooperation. In chapter 1, I develop Elgin's (2017) insight that the normative status of justificatory norms arises from the deliberation of a group of idealized agents concerned with epistemic agency and responsibility, in addition to truth. Scientists sometimes deploy what Elgin calls "felicitous falsehoods" in their models to better understand complex features of the world. So, for an epistemic achievement in science, truth is neither necessary (because of felicitous falsehoods) nor sufficient (because of the requirement of significance, and of truths being ultimately accepted as common knowledge), though, of course, truth is important. In chapter 1, I develop what I call the *Leibniz procedure* to secure the foundation of a fully deontic philosophy of science.

To summarize, by deontic philosophy of science, I mean the articulation of evaluative concepts for science by reference to the satisfaction or violation of justificatory norms, which contrasts with consequentialist philosophy of science that articulates evaluative concepts for science by reference to end-states, typically the attainment of truth or knowledge. The relevant deontic principles for science are justificatory norms, some of which are strict duties or obligations, while others are looser principles of epistemic prudence.

The potential for a deontic philosophy of science to be so fruitful is partly due to the fact that consequentialist philosophy of science has been for so long under our feet without us noticing (epistemic consequentialism has also been dominant in epistemology, though Sylvan (2020) recently offered an account of epistemic nonconsequentialism). While philosophers have disagreed about what the aim of science is—empirical adequacy, knowledge, universal laws of nature, problem-solving capacity, to name some prominent examples—all seem to agree that evaluative concepts for science should be characterized in terms of reaching whatever one holds that aim to be. For example, van Fraassen (1980, 8) claims, "What the aim is determines what counts as success in the enterprise." Yet if we do not assume consequentialism about epistemic goal-seeking, this is not obviously true. Asking where the normative status of methodological rules comes from, Siegel (1990, 301) claims, "A methodological rule is justified insofar as it maximises the likelihood that experimentation conducted in accordance with it leads to true (or valid) results." Chapter 1 discusses some reasons to think that this position is incomplete. Bird (2022, 16) argues that because the aim of science is knowledge, "science is successful when it produces knowledge." I agree with Bird that the aim of science is a special kind of knowledge—another point I argue for in chapter 1—yet a deontic philosophy of science can grant that science has a constitutive aim of knowledge without requiring the achievement of knowledge as a condition of success. Laudan (1977, 13) had a completely different conception of the aim of science than Bird, but he nonetheless claimed that the "first and essential acid test for any theory" was whether "it provides satisfactory solutions to important problems." So, for Laudan, as for van Fraassen, Bird, and Siegel, and many other philosophers representing a diverse range of programmatic commitments, evaluative concepts for science should be based on whether science has attained its aim, however that aim is construed. Indeed, the entire realism-antirealism debate clearly assumes a consequentialist philosophy of science since the

INTRODUCTION   13

debate is about whether science can attain a particular aim (crudely: truth about unobservables), with realists affirming and antirealists denying. In the following chapters, I show that many other positions on specific philosophical questions about science have relied on a consequentialist philosophy of science. For a wide range of conceptual and evaluative questions about science that perhaps have begun to seem stale, a deontic philosophy of science can offer a completely fresh outlook.

You Can't Handle the Truth

In science, you can't handle the truth. You can touch this microscope, you can see that the glass tube in your hand contains a red liquid, and you can hold the guinea pig you are about to sacrifice just as Marie Curie held the uranium ore from which she extracted polonium. But the truth about that red liquid? The truth about what happens inside that guinea pig? The truth about polonium? Truth comes later. Truth cannot be held. You can't handle the truth.

Unlike many quotidian claims or beliefs, the truth of most scientific claims is not immediately epistemically accessible. I can see that there is a cup of coffee in front of me, so the proposition "there is a cup of coffee in front of me" is true, and, putting aside scenarios in which an evil demon is tricking me, I can immediately see that it is true, and I can, of course, hold the cup and drink from it—in many quotidian scenarios, I can handle the truth. But in science, that is not typically the case. Indeed, if the truth about an aspect of the world were immediately accessible, then there would be no need to study that aspect of the world scientifically.

In the previous section, I gave an argument for a deontic characterization of evaluative concepts for science based on an analogy with musical goals. The intuition that I hope I elicited suggests a very general attitude one should take when evaluating actions, namely, that judgment of action should be based on the quality of the means one takes to achieve one's end rather than on the attainment of the end itself. In the context of science, that entails articulating evaluative concepts based on the extent to which a particular scientific work satisfies justificatory norms, rather than whether a particular scientific work discovers truth. There are other powerful arguments for deontic characterizations of evaluative concepts for science, based fundamentally on the fact that in science, you can't handle the truth. In the rest of this section, I describe three such arguments that I have alluded to previously that will reappear in various chapters throughout the book.

First: Truth in science is *retrospective benediction*. That is, the ascertainment that a putative scientific finding is true takes time, often years, decades, or even centuries after a scientific claim is first articulated or receives some initial empirical warrant. For example, experiments in the 1940s suggested that genes are composed of DNA, but it was not until that claim was justified by other experimental findings over the next ten years or so that geneticists considered the claim to be true (I return to this example in chapter 6). An extreme example of retrospective truth benediction is the gradual adoption of the sun-centered model of the solar system, which took many decades to be widely deemed true by scientists. The fact that ascertainment of the truth of a scientific claim occurs later than the theoretical articulation of that claim or the production of empirical justification for that claim entails that typically the truth or falsity of a claim is not epistemically accessible when that claim is first articulated or when the initial evidence for that claim is presented. Since the truth-status of a scientific claim is not epistemically accessible in real time, truth cannot be the basis for real-time evaluation of science.

Second: Truth is a *nirvana norm* for science. A nirvana norm is a lofty ideal that provides little guidance for those to whom the ideal is meant to apply. Suppose your yoga guru tells you to seek nirvana. It is not clear what you should do next to pursue this lofty aim. Should you drink less wine (or more)? Which restaurant should you go to for dinner? Should you become an investment banker or a philosophy professor? We saw earlier that ethical codes can give clear guidance for action. The Buddhist tradition has the Noble Eightfold Path, which includes clear guidance in eight domains ("paths") of life: For example, the right livelihood path tells people not to earn a living by selling meat, alcohol, or people, while the right speech path tells people to speak the truth and to not engage in idle small talk. Truth is a nirvana norm, unlike the many justificatory norms in science, which typically provide clear guidance for science. (Laudan [1977, 1984] argued that truth is a utopian aim, one that is impossible to achieve or even know that we have achieved it. In chapter 1, I argue that Laudan was wrong about this and argue instead that truth is a nirvana norm.)

Third: Truth does not figure in *scientific sanction*. The sole basis for scientific sanction, at least on epistemic grounds, is justificatory status. If I say to you that the sky is green or that Leibniz lived in the nineteenth century, you can reject my claims by responding that they are not true. Their falsity is sufficient for you to reject my claims and perhaps to judge me negatively as a source of information in general. Yet for live science, truth or falsity is not (and cannot be) a basis for accepting or rejecting claims or evaluating the speakers of claims. That is partly because of the first argument above—namely, that truth is not epistemically accessible in science at the moment of assertion. But there is an equally important reason, central to the deontic philosophy of science developed in this book. Scientific judgment should be based on the presence or absence (or, better, degree) of justification for a scientific claim rather than on the truth or falsity of a claim. A good example is the discovery that genes are composed of DNA. Initial experiments that suggested that genes are composed of DNA had some significant potential biases, so the scientific community was cautious in interpreting this evidence in the following decade, with most geneticists believing that the finding was spurious. This negative evaluation of the finding that genes are composed of DNA was based on an insufficient initial justification. The finding was later shown to be true, but its truth-status did not and could not figure in the contemporaneous judgment of the scientific work.

Here is another example illustrating the emphasis on justification rather than truth in scientific evaluation by scientists. During the Manhattan Project, physicist Edward Teller designed the "Super," a proposed thermonuclear hydrogen bomb that would be many times more destructive than the atomic bombs being developed at the time. But subsequent calculations by others showed that the designs for the Super were flawed. Physicist Hans Bethe later wrote, "Nobody will blame Teller because the calculations of 1946 were wrong, especially because adequate computing machines were not then available. But he was blamed at Los Alamos for leading the Laboratory, and indeed the whole country, into an adventurous program on the basis of calculations which he himself must have known to have been very incomplete" (cited in Rhodes 1986, 722). Bethe articulates the blamelessness of factual error and the blameworthiness of insufficient justification. Putting aside any moral qualms you might have about developing nuclear weapons, Bethe's point is that being factually wrong is not grounds for scientific sanction, but being insufficiently justified is grounds for scientific sanction.

A final example. L'Aquila is a medieval Italian city nestled in the Apennine Mountains. One terrible night in 2009, an earthquake of 6.3 magnitude shook the fragile town, damaging thousands of buildings and killing hundreds of people. A week before, local officials and earthquake scientists had met to discuss daily tremors in the city and the possibility of a large earthquake. The earthquake specialists at the meeting correctly claimed that despite the tremors the probability of a serious earthquake remained low. Nevertheless, after the earthquake, prosecutors pressed criminal charges against the scientists, accusing them of negligence. They were found guilty of manslaughter and sentenced to six years in prison (Pamuk 2021). The prosecutor argued that the scientists had given "incomplete, imprecise, and contradictory information" (Hall 2011). Major science organizations and thousands of scientists wrote letters to the president of Italy, complaining that the basis for the charges—that the scientists did not accurately predict the earthquake—was unfair. Assessment of science should be based not on whether the conclusion of particular scientific work is true but rather on whether the work deploys adequate justificatory practices.

To summarize: In science, you can't handle the truth, and this has significant implications for our evaluative concepts for science. Rather than truth, our evaluative concepts for science should be based on justification. Justification is the heart of science and the core of a deontic philosophy of science.

Refining Deontic Philosophy of Science

One might think that a difference between deontological ethics and deontic philosophy of science is that ethical principles are plausibly timeless, while scientific principles cannot be timeless because they are discovered and developed at particular moments in history. All of the above examples are principles or methods that were articulated fairly recently—the randomized controlled trial, for example, was devised in the first half of the twentieth century by statistician Ronald Fisher. The ethical principle "don't lie" is not historically contingent—it is plausible to think that such a principle applied to our ancient ancestors even if they did not have the language or concepts to express it. But the scientific principle "perform a randomized trial if one wants to test a drug" was devised only one century ago. However, I do not think this is much of a disanalogy between deontological ethics and deontic philosophy of science. One could say that when our ancestors wanted to test the effectiveness of a drug, they would have been better off using a randomized trial—whatever reasons we have for using such a method, those reasons would have applied to our ancestors as well, even if they were not aware of those reasons.

This response, though, creates a puzzle. If the reasons that ground a scientific principle are timeless, then one might think that the principle itself has normative force even if it has not yet been formulated, and if a principle has not yet been formulated, then a scientist very likely cannot satisfy it. It would follow that a scientist could do their very best yet still violate a norm; they acted one way, but they should have acted another way, and that is wrong. Some philosophers maintain that the normativity of "ought" is objective in the sense that the normative demand of an ought-claim applies regardless of whether the agent to whom the ought-claim applies is aware of its demands. Yet most philosophers accept the idea that "ought implies can," and it is only a very strained notion of "can" that affords an agent a capacity to satisfy a complex moral or epistemic principle that has not yet been articulated.

A defender of the objectivity of ought-claims might say that there are two ways an agent can fail to be aware of an ought's demands: by not knowing of the ought itself or by not knowing the facts that are relevant to instantiating the ought in a particular situation. In both cases, one might believe that a third-person perspective exists, an objective perspective, which can only be accounted for if ought is objective. Illustrating the first case: A scientist who tested a drug in the eighteenth century by consuming it herself and observing the putative effects would not have satisfied the norm of performing a randomized trial. Of course, she could not have known that this was a justificatory norm that she should adhere to, but despite her not knowing this, it would have been better if she had performed a randomized trial, and we can only make sense of that "better" if we grant that ought is objective. Illustrating the second case: A surgeon is treating a patient who displays all the signs and symptoms of a particular disease. It turns out, however, that those signs and symptoms are misleading, and, consequently, the surgeon misdiagnoses the patient and performs the wrong procedure, and so the patient dies. It would have been better if the surgeon had made the right diagnosis because the patient would have lived, and we can only make sense of that "better" if we grant that ought is objective. Thus, goes the argument, ought is objective.

Normativity, though, is about guiding action and evaluation (I am treading roughly over delicate ground; for a more delicate and necessarily lengthier treatment, see Gibbons [2013]). An agent can be guided by an ought-claim only if they are aware of the ought and the relevant facts that instantiate the ought. The scientist in the above paragraph could not have known of the norm in question since it was not developed until many decades later, and the surgeon in the above paragraph could not have diagnosed the correct disease because all of his evidence was misleading. Similar considerations apply to evaluation: When you ask yourself if the scientist or surgeon did something wrong, I believe you will say no. It follows that either you do not believe they violated a norm or you do believe they violated a norm, but you consider the norm violation excusable. What, though, would ground the excusing? Presumably the lack of knowledge of the norm or of relevant facts instantiating the norm, but that suggests that the real normativity of an ought-claim requires awareness of the ought-claim and the relevant facts that instantiate the ought-claim.

Consider another medical thought experiment, from Jackson (1991). You are a doctor treating a sick patient to whom you can give one of three drugs: Drug A will alleviate many of the symptoms, leaving the patient a little sick but much better off than without the drug. Either Drug B or Drug C will completely cure the patient while the other will kill the patient, but you do not know which is which. You must decide which drug to give the patient. You will say, I believe—along with every other person to whom I have posed this thought experiment—that you should give Drug A to the patient. The defender of the objective ought, to be consistent with the previous cases of the scientist and surgeon, must say that the best choice is to give whichever drug would completely cure the patient, regardless of your lack of knowledge of which drug that is. From this, it follows that the best choice is either Drug B or Drug C. But *that*, as all people seem to agree, is implausible. Thus, there is no objective ought.

To be clear, denying that there is an objective ought is not the same as claiming that there is no objectively right thing to do in any given situation. The question about whether there is an objective ought asks which perspective evaluations of action should be made from: the agent who must decide on an action, typically with imperfect knowledge of all relevant facts and governing principles, or a third-person onlooker who (by stipulation) has knowledge of all relevant facts and governing principles? I have argued for the former, and I hope you will agree that the latter is implausible, or at least practically idle. There is, though, an objectively right choice for an agent who must choose an action conditional on their beliefs about the relevant facts and governing principles. I cannot do justice to the extant literature on the question of whether there is an objective ought—I have already made a detour through intellectual territory that philosophers of science do not normally wander in, and a fair treatment of the extant literature would take me further still from concerns about science.

Veritism is a philosophical theory that says that the fundamental epistemic good is truth, and practices of justification are important only insofar as they promote the attainment of truth. Veritism is, then, a version of "epistemic consequentialism," and an epistemic consequentialist maintains that there is a single fundamental epistemic good and all putative justificatory norms receive their normative status only because they contribute to achieving that fundamental epistemic good (which, for veritists, is truth). Epistemic consequentialism is primarily a theory about an individual's beliefs. I close this section by describing an argument from Berker (2013) against epistemic consequentialism. Suppose Sasha has an odd quirk: Every time she wonders if a number is prime, she concludes that it is not. She wonders if 12 is prime and concludes it is not; she wonders if 15 is prime and concludes it is not; she wonders if 113 is prime and concludes (wrongly) it is not. Because the ratio of prime to nonprime numbers gets smaller as the numbers one considers get larger, the more numbers Sasha wonders about, her ratio of true to false beliefs about prime numbers increases; as the numbers she wonders about go to infinity, her proportion of beliefs about prime numbers that are true approaches one. So Sasha's method of assessing prime numbers is a reliable guide to truth, at least for sufficiently large enough numbers, and her method becomes more reliable the more she uses it. An epistemic consequentialist must say that, since a justificatory practice gets its status from its capacity to deliver true beliefs, Sasha's method is good. If you believe there is something inappropriate about Sasha's belief-forming process, you may have some sympathy for a deontic evaluation of epistemic practices. In chapter 1, I describe similar cases and argue that they support a fully deontic philosophy of science.

In short, I propose a deontic approach to our evaluative concepts for science in which the pertinent norms must be epistemically accessible, be action-guiding, and provide a basis for evaluation. The implications of a deontic philosophy of science are wide-ranging and indeed can completely overturn some dogmas in philosophy of science, as I argue in subsequent chapters.

Alliances

Philosophers are guilty of telegraphing ideas with *-ians* and *-isms*. Our -ians are usually based on the names of famous dead men, such as Rawlsian political theory, Foucauldian genealogy of knowledge, Kantian ethics, and Aristotelian metaphysics. Our -isms are more numerous and include internalism, externalism, Platonism, liberalism, empiricism, rationalism, naturalism, realism, antirealism, pluralism, pragmatism, and many other polysyllabic monstrosities. Sometimes -ians and -isms get combined, as with humanitarianism, totalitarianism, and Bayesianism. Like all academic disciplines, philosophy uses technical terms to convey complex meanings in a compressed code. Our -ians and -isms identify packages of interrelated ideas that comprise a commitment to a particular philosophical theory, a school of thought, or a way of doing philosophy. So, for example, pragmatism means something like "our philosophical theories should make a difference to the world, and we shouldn't worry too much about abstract intellectual puzzles of no importance." I occasionally use -ians and -isms in this book. This calls for an apology since such shorthand can be a lazy way of communicating, and it risks misunderstanding—there are, for example, many kinds of "pluralism" in philosophy of science, and so when philosophers use that term to signify their commitments, they do not convey much. My general approach to philosophy is antiismianism.

Yet our -ians and -isms convey allegiances and alliances. Deontic philosophy of science can be seen as an ally of a broad range of movements in philosophy of science, epistemology, and even sociology and history of science. Though the emphasis in deontic philosophy of science involves a shift of focus for general philosophy of science, many subdisciplines in the humanistic study of science have been concerned, one way or another, with the heart of science.

Much of philosophy of science today seems unconcerned with evaluative concepts for science in general because much of the literature is, as Cartwright et al. (2022) rightly note, "particularist," concerned with fine details of cases rather than lessons about science in general, and is descriptive rather than normative—this is particularly so for the so-called philosophy of science in practice movement. At the other end of the spectra of particular-general and descriptive-normative, mainstream epistemology has, in the last generation, undergone a normative turn, moving away from dry analyses of concepts like knowledge ("S knows that p if and only if ..."), normative only in an anemic sense, while moving toward full-blooded normative topics such as the conditions for appropriate belief or assertion.

The philosophical project of this book shares the ambition of recent epistemology to be both general and normative, but it is sympathetic with the emphasis on practical details in philosophy of science. Yet those two intellectual programs—mainstream epistemology and philosophy of science—have dangers. Much epistemological theorizing is based on an extremely impoverished notion of what counts as evidence for its philosophical theories. Some epistemologists propose thought experiments and call the intuitions that they hold in response to them "data" or "linguistic evidence," and they directly use those intuitions as "evidence" for their favored epistemological theory. This is scandalous (I describe this scandal in more detail in chapter 4). Philosophers of science, on the other hand, often drown out general normative theory with their focus on fine-grained details of historical or contemporary scientific practice. Describing the anatomy of a leaf dangling on one's favorite tree cannot contribute to developing principles of sustainable forestry.

Arthur Fine (1986) defined his version of naturalism as maintaining that one's metaphysics of science should be determined by "the very same standards of evidence and inference that are employed by science itself." Deontic philosophy of science maintains a similar stance, not just for metaphysics but also for general evaluative concepts, with one significant tweak. Rephrasing Fine's statement, a first-pass prescription of deontic philosophy of science could be that one's evaluative attitude toward everything about science should be governed by the same standards of evidence and inference that are employed by science itself. The main (and very significant) tweak deontic philosophy of science adds to that first pass is that one should not read off what those standards of evidence and inference are by directly observing scientific practice, as a naive naturalist might; rather, one should allow for some idealization in the articulation of those standards, as I argue in chapter 1. Thus, a second-pass prescription of deontic philosophy of science, rephrasing Fine's statement and incorporating that tweak, would be that one's evaluative attitude toward all aspects of science should be governed by the very same standards of evidence and inference that ought to be employed by science itself. The project of this book is, in this sense, naturalist, without giving up on normative aspirations.

Another very important movement in philosophy of science today is Bayesianism, which is based on the idea that our beliefs are graded, giving us a notion of "degrees of belief" represented by probabilities, and the prescription that our inductive inferences should be guided by Bayes' theorem, a simple bit of math that has been used to develop profound results in philosophy and statistics. Though it is a broad church, its focus on degrees of belief rather than all-out belief, its emphasis on confirmation rather than on truth, and its underlying fallibilism and epistemic humility entail that Bayesianism should be sympathetic to a deontic philosophy of science (see Sprenger and Hartmann 2019).

Even scholarly disciplines with a history of mutual suspicion or even mutual hostility toward philosophy of science might be more amenable to deontic philosophy of science. For example, so-called third-wave sociologists of science Collins and Evans (2017) accept the message of second-wave sociology of science, which claims that science cannot attain truths, while nonetheless arguing that democracy needs science because science is a source of values. They claim that scientists "have only to try, not to succeed, to be doing good" (2017, 25); by *succeed*, they mean the attainment of useful truths; by *trying*, they mean the appropriate deployment of science's many justificatory norms. A deontic philosophy of science agrees that scientists do not need to successfully attain truths to be judged positively, yet a deontic philosophy of science can easily grant that science at its best achieves truth (in chapter 8, I argue that science at its best attains what I call "timeless truths").

A deontic philosophy of science also makes sense of how historians of science sometimes consider past science. For example, when discussing ancient Greek science, Toulmin and Goodfield (1961, 126–27) claim, "In judging them as scientists—as rational interpreters of Nature, that is—the important thing, surely, is not to ask how many conclusions they reached which we still accept, but rather how far their conclusions were supported by the evidence then available." Indeed, it is a platitude among professional historians of science that history should not be "whiggish"; that is, historians should not use present views about truth or falsity as evaluative standards for past science (Jardine 2003). A deontic philosophy of science agrees, holding that our evaluative standards for science should be based on justification rather than truth.

One of the central Confucian texts, the *Great Learning*, describes a set of eight fundamental aims, articulated as an ordered list in which subsequent aims depend on prior aims:

Investigate things, then extend knowledge;

Extend knowledge, then clarify one's will;

Clarify one's will, then set one's heart right;

Set one's heart right, then cultivate one's person;

Cultivate one's person, then harmonize one's family;

Harmonize one's family, then manage the state;

Manage the state, then achieve world peace.

One interpretation of this text holds that the antecedent in each statement can be considered a necessary condition for the attainment of the consequent (Li 2023); so, for example, to cultivate one's person, one must first set one's heart right. Especially striking is the placement of the imperative to *investigate*, the foundation for all the other lofty aims that follow. Achieve world peace by studying! The very first word in Confucius's description of the fundamental aims, the bedrock and necessary condition for all others, including world peace, is to engage in the heart of science.

Summary

In chapters 1, 2, and 3, I lay the foundations of my deontic philosophy of science. In chapter 1, I develop an account of the aim of science that I started in a recently published article (Stegenga and Menon 2023). I argue that a constitutive aim of science is common knowledge, which is a form of strong consensus in which parties to the consensus not only agree about a scientific claim but also agree that the justificatory reasons others have for their commitment to the consensus are indeed justificatory. This constitutive aim of science entails a discursive requirement for scientific practice, namely, that justificatory reasons for scientific claims must be articulated, which then exposes those reasons to critique and possibly refinement and ultimately acceptance or rejection by others. In chapter 1, I also describe the notion of justificatory norms—the domain-general principles and domain-specific practices that ultimately form the basis of claims to common knowledge—and argue that they receive their normative status as deliverance by a group of ideal deliberators who ask which norms maximize epistemic agency, epistemic responsibility, and, of course, access to important truths.

In chapter 2, I turn to a long-standing topic in philosophy of science: the demarcation of science from other nonscientific or pseudoscientific endeavors. The most famous attempt to offer a demarcation criterion for science was Popper's principle of falsifiability. While this demarcation principle became wildly popular with scientists, most philosophers rejected it long ago. This demarcation principle and others offered by philosophers of science typically appeal to a single, ungradable criterion. This is odd. It is like assessing students by a single factor, say, their classroom attendance, concluding that all and only students with perfect attendance pass while the remaining students fail. A deontic philosophy of science rejects the supposition that the demarcation of science must be based on a unitary and ungraded feature and thereby offers a fresh approach to demarcation and related notions of objectivity, bias, and trust in science.

The value-free ideal in science says that values should not influence scientific reasoning. Philosophers have criticized the value-free ideal, arguing that it is both unattainable and undesirable. The supposed untenability of the value-free ideal has nearly become dogma among philosophers of science. In chapter 3, I offer a new version of the value-free ideal and argue that it can be defended as a principle for science even if the unattainability and undesirability of a value-free end-state are granted. If a goal is unattainable, then one can separate the desirability of accomplishing the goal from the desirability of pursuing it. Developed with Tarun Menon, this novel value-free ideal holds that scientists should act as if science should be value-free (Menon and Stegenga 2023). We argue that even if a purely value-free science is undesirable, this value-free ideal is desirable to pursue, thereby demonstrating again the attraction of a justification-oriented deontic philosophy of science.

In chapter 4, I ask what norms scientific assertion should be held to. Scientific assertions include any claim about the world in any communicative context, such as technical journal articles, public-facing editorials, or reports to policy-makers. The most popular account of assertion in general today holds that assertions are governed by a knowledge norm—that account thus posits that for an assertion to be appropriate, it must be known by the person making the assertion. Against this, I argue that scientific assertions should be justified and informative. I make several arguments against factive norms of assertion in science: Scientific assertions are not governed by a truth norm or a knowledge norm. I then illustrate the violation of informativeness and justification by two prominent scientific research programs during the COVID-19 pandemic.

In chapter 5 I ask, What is scientific progress? I defend a novel account of scientific progress centered around justification. Science progresses, on my account, when there is a change in justification. This account of scientific progress challenges the idea that scientific progress requires an accumulation of truth or truthlikeness, and it emphasizes the social nature of scientific justification, drawing on my account of common knowledge from chapter 1.

Many philosophers who study the allocation of credit in science seem to believe that the so-called priority rule—which holds that the first scientist to discover something gets all the credit for that discovery—is the principle by which credit in science is allocated. Using technical models, philosophers have noted various consequences of the priority rule, such as efficient allocation of research resources or, conversely, hasty, nonreproducible research. A substantial edifice of scholarly work is predicated on the primacy of the priority rule. In chapter 6, I argue that the priority rule is a descriptively inaccurate and normatively inappropriate account of the reward system in science, and I offer an alternative account of scientific credit based on justification, again illustrating the novel potential of a deontic philosophy of science.

In a supreme emergency, such as the recent COVID-19 pandemic, scientists might violate some of the justificatory norms of routine science to quickly develop interventions against the catastrophic threat of that emergency. In chapter 7, I call this "fast science." The magnitude, imminence, and plausibility of a threat justify engaging in and acting on fast science. Yet that justification is incomplete. I defend two principles to assess fast science: (1) fast science should satisfy as much as possible the justificatory norms of routine science, and (2) fast science developing an intervention against a threat should not depend on the same norm violations as the fast science that estimates the magnitude, imminence, and plausibility of the threat.

Many scholars and even some scientists appear uncomfortable with the idea that science can achieve what I call "timeless truths." In chapter 8, I address the provisionality thesis, mentioned above, which holds that in science there are no timeless truths because all scientific findings are liable to be overturned in the future. I argue, in contrast, that science can attain timeless truths, or what Vickers (2023) calls "future-proof science," because science can attain common knowledge. Simple examples abound: Genes are composed of DNA, the structure of DNA is a double helix, water is a compound of two hydrogen atoms and one oxygen atom, and the sun is a star. Because the conditions for attaining common knowledge are very demanding, as described in chapter 1, when those conditions are met and thus when science attains common knowledge, such knowledge is not liable to be overturned by future science. Rather, common knowledge is composed of timeless truths.

As Woodward (2003) notes at the beginning of his book on causal explanation, writers can become focused on a single idea, try to push it as far as it can go, and end up claiming for it more than is warranted. I am probably guilty of this. Nevertheless, I believe it is worthwhile to explore the philosophical implications that can result from a focus on the heart of science.

CHAPTER ONE

Common Knowledge

1.1 Introduction

The aim of science is common knowledge. Common knowledge is a special kind of knowledge in which interested agents form a consensus about a proposition and its justification. Suppose a small community of two scientists, Sasha and Masha, both justifiably believe a true finding, x. They have a weak consensus about x, but they do not yet have common knowledge; both Sasha and Masha must articulate their respective justifying reasons for believing x. After critical scrutiny, Sasha may come to accept Masha's justification, and Masha may come to accept Sasha's justification—at that point, x becomes common knowledge for this community.

In §1.2, I argue that common knowledge is the aim of science. In §1.3, I refine the definition of common knowledge and develop the argument that common knowledge is the aim of science. What distinguishes common knowledge from simple knowledge is its social nature and particularly the requirement for consensus about justification. In §1.4, I give an account of the source of normativity for justificatory norms, which is based on what I call the *Leibniz procedure*. In §1.5, I respond to some possible objections to the main theses developed in this chapter, and in §1.6, I briefly put the notion of common knowledge in a broader context and articulate its relation to a range of other positions in philosophy of science.

1.2 The Aim of Science

Science is, of course, a large and complex institution with a huge number of stakeholders, most prominently scientists, the majority of whom have a range of interests that go beyond discovering truths. Some of these interests include having a stable career, doing fun things in the laboratory or the field, and occasionally earning the grand forms of credit that science awards, such as prizes. So the aim of science cannot be a function of the aims of the individual stakeholders in science. Rather, the aim of science is a constitutive aim of the institution itself, just as the aim of a corporation is profit, the aim of a judicial system is justice, and the aim of chess is victory. In this chapter, I argue that the constitutive aim of science is common knowledge.

How can one identify the constitutive aim of an institution? I describe four kinds of considerations one can appeal to in order to address this question.

First, if a postulated aim of an institution can explain the existence of the many diverse practices of that institution, that is some evidence that the postulated aim is in fact the constitutive aim of that institution. For example, we can identify the constitutive aim of medicine as caring for and curing sick people because that is what most practices in medicine are supposed to contribute to.

Second, if an institution having a particular constitutive aim would create benefits that would not otherwise occur if the institution did not have that aim, that would be a defeasible consideration that the institution does indeed have that constitutive aim. For example, the fact that chess has the aim of checkmating one's opponent solves a coordination problem that generally goes unnoticed because its solution is so successful, that is, how to ensure that both players have the same general idea about how to behave when playing a game of chess. Consider what playing chess would be like if it did not have that constitutive aim. Suppose, for example, that the aim of chess was a convoluted disjunction, like "If you are a Scorpio, identify the most beautiful chess piece on the board; if you are a Libra, juggle the most number of chess pieces that you can; if you are a Capricorn, play checkers with the chess pieces," and so on for the other horoscope signs. Then, if you were to play a chess game with a random opponent, you would not know what to expect from them or how to interact with them via the structure of the game—it would be much like playing chess with a three-year-old or a puppy. (That is not to say, of course, that all people who play chess are motivated, entirely or in part, by victory, though many certainly are. One's goal in playing chess may simply be to have fun or spend time with a friend. The constitutive aim of an activity or institution is not determined by the goals of particular people involved in that activity or institution.)

Third, if a postulated aim of an institution accords with our conceptual understanding of the aim, the institution, and the relation between the two, that is some evidence that the postulated aim is in fact the constitutive aim of that institution, and if there is discord between our conceptual understanding of the aim, the institution, and their relation, that is evidence that this postulated aim is not the constitutive aim of that institution. For example, if one claimed that the aim of judicial systems is to punish criminals, a plausible response would be that this represents a limited understanding of the proper aim of judicial systems because one way to achieve that aim would be to punish every suspect no matter how little evidence exists that they had committed a crime. One's aversion to such a system can be understood by recognizing that there is another more fundamental aim of judicial systems, and it is that more fundamental aim that is the constitutive aim of judicial systems.

Fourth, if an institution has a constitutive aim, then once an instance of that aim is reached, we should not expect to see any further pursuit of the aim in that instance. For example, a game of chess is over when one player wins (or when no player can continue, resulting in a draw). Once a player wins, it would be absurd for either player to keep trying to win the game, which suggests that the constitutive aim of chess is victory. (This consideration applies to institutions with bounded ends, such as chess's bounded end of victory, but it cannot apply to institutions with unbounded ends, such as a corporation's profit.)

A scientific claim is common knowledge if and only if the claim is true and there is consensus about the claim and the respective justifications of the claim in the relevant scientific community. Further, that consensus must be reached by critical scrutiny of the respective justifications. Each of the above four kinds of considerations suggests that the constitutive aim of science is common knowledge.

First, many practices in science are designed so that scientific findings can be candidate claims to common knowledge. Justificatory practices of science serve to render scientific claims as justified as possible, thereby giving them at least a candidate status to knowledge. Moreover, these justificatory practices are typically public—as Gerken (2022) argues, justifying reasons in science should be publicly articulated, and science has many venues for such public articulation of justification, including the methods and results sections of scientific articles, conferences, and less formal kinds of interscientific communication such as laboratory meetings. When justification is made public, other members of the scientific community can assess the putative justifying reasons and agree with or criticize those reasons. This, in turn, renders any resulting consensus about a scientific claim a candidate for common knowledge. As Staley and Cobb (2011, 479) rightly note, a central part of science is "rationally persuasive communication in which reasons are presented to other members of the community that will serve to underwrite, within that community, the status of particular claims as knowledge." A compelling explanation for why these practices are so widespread is that a constitutive aim of science is common knowledge. Conversely, if a scientific practice hinders the pursuit of the aim of common knowledge, then it is criticized and ultimately rejected. For example, publication bias in medical research has been a widespread practice, but critics argue that it steers the community away from the truth. Publication bias clearly hinders the public sharing of justification, rendering scientific findings in this domain poor candidates to common knowledge. As a result, the community now attempts to mitigate publication bias. This can all be accounted for by the fact that the constitutive aim of science is common knowledge.

Second, the fact that common knowledge is the constitutive aim of science brings many benefits. The achievement of knowledge itself, common or not, is itself a benefit. Characterizations of the value of knowledge differ. Some say truth itself is intrinsically valuable (for example, Bishop and Trout 2005, 97), and truth is necessary for knowledge. Others argue that only significant truths are valuable (Kitcher 2001), where significance arises from curiosity or practical importance. Still others argue that knowledge is valuable because it stabilizes social coordination (Chrisman 2022). Dogramaci (2015, 796), for instance, claims that "we use epistemic evaluations to promote coordination, which is valuable because it helps us to pursue true belief as a team, a team of parallel epistemic processors who can safely share their results through trustworthy testimony." Epistemic evaluation of a claim involves assessing the reliability of the rules or norms used to justify the claim, and that is important because it promotes interpersonal cooperation (see also Dethier 2023). When a claim to knowledge is *common* knowledge, that claim is particularly good at promoting interpersonal coordination. Thus, the aim of common knowledge directly confers distinct benefits insofar as the achievement of common knowledge itself has benefits. The fact that common knowledge is the aim of science also confers many indirect benefits, which accrue thanks to the *striving* for the aim rather than its achievement. Such a benefit was discussed above: To achieve common knowledge, scientists must publicly articulate reasons for their assertions, and this in turn affords critical scrutiny of those assertions and reasons, which is the foundation for their possible claim to objectivity (Longino 1990). Therefore, the pursuit of the common aspect of common knowledge contributes to the achievement of knowledge itself, and that in turn forms the basis of subsequent scientific work, the refinement of justificatory practices, and future claims to knowledge.

Third, the fact that common knowledge is the aim of science accords with how best to understand science as an institution. There is a conceptual link between science as an institution and both the *common* and the *knowledge* aspects of common knowledge. First, let us consider knowledge. Though there has been much debate about how to properly understand the concept of knowledge, most philosophers agree that knowledge requires truth and justification. Therefore, if a person knows x, or if we speak about scientific knowledge about x, most philosophers hold that x must be true and justified (perhaps in addition to other conditions). Bird (2022) convincingly argues that science does not merely aim at truth because accidentally discovered truths or unwarranted truths are not fully scientific. Science, Bird argues, aims at *justified* truths. Thus, the aim of science is (a kind of) knowledge. Now consider the common aspect of common knowledge. This requires establishing that the constitutive aim of science is not reached until relevant agents to a consensus endorse the other agents' justifications for the object of consensus. Much of scientific practice is tuned to this. For example, Brown (2020, 40) argues that science is concerned with "public claims, which are initially warranted by inquiry and over time gain credibility through a public and social process." The gaining of credibility through a public and social process is the basis of a claim to common knowledge. In §1.3, I expand on this consideration, arguing that the common aspect of common knowledge is more robustly social than the brief remarks here suggest.

Fourth, when a particular scientific claim is admitted as common knowledge, scientists no longer research that claim—or, at least, they do not do so at the same level of granular detail. Consider an established contribution to common knowledge, such as the double-helical structure of DNA. Once that finding was admitted as common knowledge, inquiry about that claim ended and scientists moved on to study other things. No scientist after 1965, say, studied the macrostructure of DNA because that had already been, for a decade at least, a contribution to common knowledge (scientists today continue to study the structure of DNA, but they are studying finer-grained structural details and not the double-helical macrostructure). This consideration suggests that the constitutive aim of science is common knowledge. Conversely, we can ask about the conditions under which we consider an endeavor to be unfinished or incomplete. If you are playing a game of chess that is about a dozen moves from a checkmate and your game is interrupted, you are likely to feel a sense of dissatisfaction (you might feel relief if you were on the losing side, though I reckon even then you might feel dissatisfied). In science, if a putative claim to knowledge is not widely accepted, if the various justifications for that claim are not widely accepted, or if we simply do not understand some particular question, scientists are compelled to continue working on the question, suggesting that they view the matter as unfinished. Once that claim is admitted as common knowledge, there is no further impetus to work on it, though, of course, there may be impetus to investigate further downstream questions or questions at different scales or to pursue technological applications of the finding.

So the constitutive aim of science is common knowledge. One final refinement of this thesis is required to address Kitcher's (1993) compelling claim that the aim of science is not merely truth but rather significant truth. The notion of common knowledge has a dimension of significance built in, as it were, since consensus about a scientific claim could only come about if a broad swath of a scientific community cared about it.

The printing press was a technological development that facilitated the achievement of common knowledge to such a degree that the spread of this technology was highly correlated with the advance of science. According to Toulmin and Goodfield (1961, 183–84):

> It would be hard to exaggerate this change in the efficiency of scientific communication. Whereas the pace of scientific advance had depended earlier largely on the concentration of scholars in one and the same place, from now on men living at a distance from one another could collaborate effectively. Before 1500, one can hardly speak of mathematical or scientific advances ever becoming 'common knowledge': important insights were achieved at one place and time, only to be lost again in the next century, and scholars working in different cities took as their starting-points quite different bodies of knowledge.

Though these historians use the term *common knowledge* in a colloquial sense, their point stands if we substitute my technical sense of the term. The printing press facilitated the rapid sharing of scholarly work, and the public sharing of scholarly work is a precondition for the achievement of common knowledge. So the correlation between the spread of the printing press and the growth of science is no mere correlation; indeed, the relation is not merely causal, either: The ability to share results and justifying reasons is partly constitutive of the aim of science. The notion of common knowledge reaches back to embrace the old enlightenment project of Leibniz (1678–79), who believed that the well-being of humanity could be improved by developing an encyclopedia of an "orderly collection of truths" that "will be like a public treasury to which could be added all remarkable discoveries and observations." Common knowledge is a public treasury, and enriching this public treasury is the aim of science.

1.3 Consensus and Scientific Knowledge

Lewis (1969) introduced a notion of common knowledge to philosophy in his study of conventions, and Heal (1978) gave an updated analysis of common knowledge. The notion has since been studied by epistemologists who debate whether common knowledge is possible (for example, Lederman [2018], who entertains the odd thesis that common knowledge is not possible, and Immerman [2022], who, rightly in my view, argues that common knowledge is indeed possible). Philosophers of science have emphasized the social nature of scientific knowledge (Longino 2002; Bird 2022), yet, as far as I know, the notion of common knowledge has not thus far been significant for philosophy of science.

Much of the literature in philosophy of science that addresses the epistemic role of consensus in science has focused on the question of whether and when one can use consensus as a reliable indicator of truth. For instance, can we regard the consensus among climate scientists about anthropogenic global warming as a good reason to believe that Earth is in fact being warmed by human activity (Oreskes 2004)? Some argue that the consensus should not be regarded as a good reason for belief because the existence of the consensus can be explained not by epistemically relevant factors but by sociological, political, or economic factors (such as the lack of grant money for dissenting views or hierarchically maintained groupthink within the scientific community). As a consequence, philosophers have attempted to delineate the conditions under which a consensus is legitimately *knowledge-based* (the term comes from Miller [2013]; see also Tucker [2003]). The epistemic significance of consensus is often characterized as an indicator of scientific achievement and not constitutive of scientific achievement. As Lynch (2019) claims: "Agreement is just a sign that we may be closer to the truth." On this view, science need not aim for consensus itself but rather should aim for its epistemic end, which has been characterized in a variety of ways—as truth or truthlikeness (Niiniluoto 2014), knowledge (Bird 2022), understanding (Potochnik 2017), or empirical adequacy (van Fraassen 1980)—and any resulting consensus is merely an indicator of epistemic success. Vickers (2023) has recently argued, for instance, that if a scientific community is sufficiently diverse and reaches a consensus about a claim, nonexperts can appeal to that consensus as a sign that the claim is true. A consensus that emerges as a consequence of a search for knowledge could be knowledge-based, on this view, but the consensus is a by-product of the achievement of knowledge rather than constitutive of that achievement.

This position on the value of consensus is incomplete. Consensus should be construed as partially constituting the aim of science. It is not an independent aim of science; the only consensus worth having is (in a sense to be explained) knowledge-based. However, this does not mean that the importance of scientific consensus reduces to the importance of scientific knowledge because scientific knowledge is (again, in a sense to be explained) consensus-based. Neither knowledge nor consensus is a mere by-product of the other aim, but they are not separate aims either. Rather, in science, knowledge and consensus are deeply interwoven, making it impossible to distinguish between the pursuit of knowledge and the pursuit of consensus. Epistemically legitimate consensus-forming procedures are important precisely because they expand our capacity to acquire scientific knowledge.

Strong Consensus

In the domain of public reason, theorists such as Habermas (1996) argue that consensus must be achieved not merely by a unanimously held conclusion but by arguments for that conclusion that are endorsed by parties to the consensus. Scientific knowledge requires a similar type of strong consensus. Of course, requiring every member of a community to endorse a conclusion and the mutual reasons for the conclusion is unrealistic. We can conceive of a two-dimensional spectrum, with one dimension representing the proportion of a relevant community assenting to a conclusion (one end of this dimension is total unanimity about a conclusion, while the other end is total disagreement) and the other dimension representing the proportion of a community assenting to the various justifications for that conclusion offered by other members of the community (one end of this dimension is a total endorsement of others' justifying reasons for the conclusion, while the other end is total rejection of others' justifying reasons)—weak consensus requires a high value on the first dimension only, while strong consensus requires a high value on both dimensions.

Strong consensus, then, involves more than mutual knowledge. If every member of a scientific community knows C, it does not follow that there is a strong consensus regarding C. Suppose two scientists, Aisha and Bheem, rely on the same evidence and arrive at the same conclusion, C, yet they rely on different sets of justificatory norms, and both sets of justificatory norms are actually justifying, so that both Aisha and Bheem know C. Suppose, however, that Aisha does not endorse Bheem's putative justificatory norms as genuinely justificatory, and vice versa, so Aisha does not believe that Bheem knows C, and Bheem does not believe that Aisha knows C. This is a case of mutual knowledge, where both parties know C, but neither party knows that the other party knows C. This is a weak consensus.

For there to be a knowledge-based strong consensus, there must be common knowledge. In other words, not only must it be the case that both Aisha and Bheem know C, it must also be the case that Aisha knows that Bheem knows C (and mutatis mutandis for Bheem), that Bheem knows that Aisha knows that Bheem knows C (and mutatis mutandis for Aisha), and so on. Typically, this condition will be met when Aisha and Bheem have common knowledge that each of their sets of justificatory norms is genuinely justificatory and sufficient for knowledge. Strong consensus does not require that they rely on the same justificatory norms—and thus maintaining that common knowledge is the constitutive aim of science is perfectly compatible with methodological pluralism for science—but it does require that each party to the consensus has the in-principle capacity to evaluate, criticize, and endorse the other's justificatory norms. Strong consensus about a scientific claim is only possible if there is consensus about the genuine normative status of the justificatory norms used to justify the claim (a similar point is made in Lehrer and Wagner [1981]). I will call this the requirement of *shared justificatory norms*.

I use *strong consensus* and *weak consensus* as technical terms with the meanings just stated. One might think that the term strong consensus implies something like "broad agreement" and that weak consensus implies less agreement. I find the former connotation of consensus redundant and the latter contradictory since consensus already implies broad agreement. In any case, the term is used in various ways in common discourse, and

I do not intend to follow common usage here. (The literature on *public reason* also makes a distinction between strong and weak consensus, and I depart from that usage here; see Quong [2011]).

One simple reason a putative claim to scientific knowledge can fail to be common knowledge is if there is not even a weak consensus about the claim, let alone a strong consensus. For example, there is not a weak consensus about the Hubble constant now, as there are multiple values of it currently considered plausible by experts. Thus, current estimates of the Hubble constant cannot now be contributions to common knowledge. Another way a putative claim to scientific knowledge can fail to be common knowledge is if there is a weak but not a strong consensus about the claim. For example, in the early 1950s, a consensus emerged that genes are composed of DNA. Evidence for this had been generated by experiments with both viruses and bacteria, yet many scientists who were convinced by the results of the former doubted the results of the latter, and thus there was, for a time, weak consensus, though a strong consensus about the claim soon emerged (I describe this case in more detail in chapter 6).

Shared Justificatory Norms

Shared justificatory norms are a requirement for knowledge-based consensus and for scientific findings to be common knowledge.

Suppose Sara has an uncanny and reliable instinct for discovering physical laws. She can observe a physical system and simply intuit the laws that apply. This intuition comes with a deep sense of conviction and is always correct. On some conceptions of justification, Sara's intuitions count as knowledge. However, such intuitions cannot plausibly count as common knowledge because the inferential route Sara uses to get from evidence to conclusion cannot be shared with, understood by, or endorsed by other members of the scientific community. They may be able to use Sara's inferences as a guide to discovering physical laws, but they cannot replicate her inferences, nor can they understand how the method works. Thus, they would not endorse it, much like Galileo's critics did not at first endorse the evidence from his telescope. Dogramaci (2015) argues convincingly that, in the context of deduction, examples like this involve an absence of rationality or justification, and this is even more plausibly true for inductive practices. Dogramaci's example is compelling: He asks us to consider Pierre de Fermat, who wrote a note in the margin of one of his books saying that he had proved what we now call Fermat's Last Theorem. Dogramaci asks us to suppose that Fermat inferred, from a few simple axioms, the conclusion of the theorem, with a big inferential leap involving no intermediate steps. Thanks to the famous proof by Andrew Wiles in the 1990s, we now know that Fermat's leap was truth-conducive, but, argues Dogramaci, it was not rational (in my terms, it was not sufficiently justified).

There is a publicity requirement on scientific knowledge (Merton [1942] 1973; Longino 1990), and particularly on practices of scientific justification (Gerken 2022). Scientific knowledge must be supported by justificatory norms that can, in principle at least, be shared, understood, and mutually endorsed.

Weak consensus in science implies that some scientists doubt the justification of a claim by their peers, even though all agree about the claim itself. That alone is a concern that motivates a resolution. Another way of putting this is that weak consensus implies dissent about the putative justificatory norms used by peers, so anything less than a strong consensus implies some dissent about an important aspect of science. For knowledge to count as *common* knowledge, it must be justified by norms that are widely accepted in the relevant community as being genuinely justificatory. This is the sense in which scientific consensus and scientific knowledge are intertwined as aims of science. Scientific consensus must be knowledge-based—specifically, it should be explained by common knowledge acquired by a high enough proportion of a particular community, and, conversely, scientific knowledge must be consensus-based—for it to count as properly scientific, it must be supported by shared justificatory norms.

As we saw above, strong consensus in science can be construed as gradable. Instances of strong consensus in practice can be achieved with less than perfectly shared justificatory norms. Suppose Aisha and Bheem both rely on very similar models to calculate some physical quantity. Their models are identical to each other except for the value of one parameter, about which they have a minor disagreement. Furthermore, neither can endorse the parameter value used by the other. Fortunately, the outputs of both their models are nearly insensitive to the value of the contentious parameter, and so their estimation of the physical quantity, as calculated by both models, is effectively the same up to their desired level of approximation. Let us assume that the predictions of both models are accurate and reliable, and this counts as knowledge. The degree of disagreement about the relevant justificatory norm in question is small enough that there might be an approximate strong consensus—good enough, perhaps, to license a contribution to common knowledge. On the two-dimensional spectrum view of strong consensus described above, the degree of agreement in this case is high on both dimensions. On the other hand, if Aisha was using a completely different model from Bheem, and they could not endorse even the basic structure of each other's models, agreement on the predictions would be a clear case of weak consensus and thus not common knowledge.

We saw above that one dimension of the gradability of consensus is the proportion of the relevant scientific community that assents to a finding. Demanding full unanimity for a claim to count as common knowledge might seem excessively stringent. I wager, however, that for many of the quotidian examples of common knowledge I have mentioned—the double-helical structure of DNA or the sun-centered model of the solar system—the relevant scientific communities do in fact have near unanimity (see Vickers et al. 2024).

Common knowledge requires (perhaps an approximate) strong consensus. So we now have a minimal sense in which consensus is an aim of science. Without consensus about justificatory norms, there can be no common knowledge. Insofar as common knowledge is the constitutive aim of science, scientists need to aim for consensus about justificatory norms. But one might think that consensus should be an explicit aim of science only in this restricted sense. Aisha and Bheem must aim at consensus about their justificatory norms, but once they have shared justificatory norms, any further scientific progress simply requires them to use these justificatory norms appropriately. There might seem to be no further need to explicitly aim for consensus once this base level of consensus about justificatory norms is acquired. Any further knowledge-based consensus will simply be a by-product of aiming at knowledge. But this ignores a crucial point about science. Scientific progress is not merely the acquisition of scientific knowledge with existing justificatory norms (see chapters 3 and 5). It is also about expanding the scope of what could be potential scientific knowledge. Justificatory norms in science are not static. A crucial part of scientific inquiry is developing new tools that allow us to acquire knowledge that has thus far been inaccessible, convert claims that fall short of status as common knowledge into common knowledge, and determine the precise limits of common knowledge. Aiming at consensus expands the domain of common knowledge, as I explain in the next subsection.

First, however, I must address an important question: What is the relevant community within which strong consensus is required for a scientific finding to be admitted to the stock of common knowledge? The term *common knowledge* can give the wrong impression that the relevant community is simply all people, similar to how the term *common sense* is used. But assessing the justificatory status of most scientific findings requires arcane knowledge and technical skills that require years of training to acquire. This suggests that the relevant community is the set of experts in the domain of the scientific finding. Of course, one can acquire a degree of expertise without formal training, such as a person with high blood pressure who reads a great deal about treatment for lowering blood pressure; one can be a metalevel expert without first-order scientific training, such as many philosophers of science; one can acquire partial scientific expertise, such as a person who completes only half of a PhD program; and one's formal expertise in one area can provide partial expertise in a cognate area, such as a professor of biochemistry who thereby also knows much about molecular biology. So the precise boundary of the relevant community within which strong consensus is required for a scientific finding to be admitted to the stock of common knowledge may not be perfectly sharp, but the relevant community is composed for the most part of the relevant scientific experts.

Consensus-Based Knowledge

Suppose Aisha and Bheem disagree about the conclusion they have reached based on their evidence E. Aisha infers from E to C_1, Bheem from E to C_2, and C_1 and C_2 are incompatible. Neither has made a mistake in their use of justificatory norms or failed to use a relevant justificatory norm, but the current set of shared justificatory norms is insufficient to justify a unique inference from E to either C_1 or C_2. In terms familiar to philosophers of science, E underdetermines the choice between C_1 and C_2. The resolution of this disagreement is a goal the scientific community is committed to qua scientific community.

There are several ways this resolution could occur. Aisha and Bheem might collect more evidence, which might settle the dispute between C_1 and C_2 without any change in justificatory norms because now the evidence is sufficient to justify a unique inference using shared justificatory norms. Or they might come to a consensus about a new shared justificatory norm that, when used, settles the dispute to the satisfaction of both scientists in favor of either C_1 or C_2. I explore these possibilities in greater detail in chapter 3 to defend a new value-free ideal for science.

Here is an illustrative example. In the 1950s, statistician Ronald Fisher (1958) denied that smoking causes lung cancer. This may have been a tenable position when Fisher published a letter in *Nature* arguing that due to the possibility of a confounding factor in the case-control studies that showed a correlation between smoking and lung cancer, we could not reliably conclude that smoking causes lung cancer. Fisher alluded to twin studies that showed that the smoking habits of monozygotic twins were more similar to each other than those of dizygotic twins, suggesting a genetic basis for smoking, and he concluded that his disputants who believed smoking caused lung cancer were value-motivated. He called their proclamations "propaganda" ("there is nothing to stop those who greatly desire it from believing that lung cancer is caused by smoking cigarettes"). In response, epidemiologist Bradford Hill (1965) argued that multiple considerations could together justify a causal conclusion in this context, including the observed strength of the correlation, the reproducibility of the findings, the biological plausibility of the causal relationship, and several others—these are sometimes called "Hill's criteria." By appealing to these considerations, it was much more challenging to deny that smoking causes lung cancer. Moreover, Hill's methodological innovation was subsequently used to advance scientific knowledge about a range of other questions.

In addition to collecting more or better evidence, it is also possible that Aisha and Bheem might come to a consensus that does not settle the issue in favor of either C_1 or C_2 but involves the identification of a new conclusion, C_3, which is the most informative conclusion entailed by both C_1 and C_2 and can be inferred from E using shared justificatory norms. In the remainder of this section, I develop this possibility in more detail.

To arrive at C_3, Aisha and Bheem do not need new evidence—C_3 was already a part of both of their individual sets of beliefs, at least implicitly, since both C_1 and C_2 entail C_3. Neither C_1 nor C_2 was a candidate for common knowledge since neither conclusion could be justified using only shared justificatory norms. Yet C_3 is a candidate for common knowledge. Aisha and Bheem have a claim that does not state anything new about the world but rather states what can be justified using their shared justificatory norms. This is not knowledge-driven consensus—consensus that is a by-product of acquiring new knowledge (or potential knowledge). It is consensus-driven knowledge. Aisha and Bheem have added to the public treasury of common knowledge by pursuing consensus.

As an example, Aisha and Bheem might both infer distinct causal models from a given set of evidence, with the difference in models due to the fact that some of their justificatory norms are not shared. To resolve the disagreement, they develop a statistical technique that they both endorse, which allows for an inference from the evidence to a more coarse-grained causal model, one for which both of their earlier models are different fine-grained specifications. The coarse-grained model specifies just those aspects of each of the fine-grained models that could be regarded as common knowledge. The details from each fine-grained model that are not included in the coarse-grained model are excluded from consideration as present common knowledge, at least until some future change in the available evidence or justificatory norms. (Bradley, Dietrich, and List [2014] have developed a formal approach along these lines.)

Longino (1990, 222) made a similar point when discussing responses to disagreement: "Whether we are reading an instrument or observing a troop of baboons, there is always some minimal level of description of the common world to which we can retreat when our initial descriptions of what is the same state of affairs differs." (Parker [2014] makes a similar point.) An example of this type of innovation is the set of tools that has been developed by the Intergovernmental Panel on Climate Change to characterize the nature and extent of consensus in climate science. These tools include ways to measure and express uncertainty that take into account both quantitative and qualitative variation in beliefs across the community. They also include multimodel ensemble techniques that involve looking at distributions of projections emerging from multiple different and incompatible climate models, or identifying outcomes that are robust across multiple models, rather than focusing on the individual projections of each model (Winsberg 2018).

So, common knowledge is a constitutive aim of science, and it requires a strong consensus in which scientists have shared justificatory norms. Scientists in a strong consensus must endorse each other's justificatory norms (such endorsement may be explicit or implicit). This raises a fundamental question: On what basis should scientists give or withhold such endorsement? An answer to this question must articulate the fundamental basis of normativity for putative justificatory principles and practices, to which I now turn.

1.4 Normativity of Justificatory Norms

Recall from the introduction Sasha's method for assessing whether a number is prime: Every time Sasha considers whether a number is prime, she automatically and unreflectively concludes that it is not prime. This is not a method that a responsible epistemic community would endorse. Suppose Masha studies the properties of the number 108341 and concludes that it is not prime. Masha asks Sasha to assess whether it is prime, and Sasha follows her usual method and concludes, of course, that it is not prime. Hence Sasha and Masha both agree that it is not prime, and thus they have at least a weak consensus that it is not prime. But Masha does not endorse Sasha's method for determining whether numbers are prime or not. So they have mutual knowledge but not common knowledge that 108341 is not prime.

In the introduction, I presented a deontic approach to philosophy of science, which involves rejecting epistemic consequentialism. However, rejecting epistemic consequentialism comes with a cost. Epistemic consequentialists have a neat and simple story about the source of the normativity of justificatory norms: A justificatory norm gets its normative status from its capacity to promote one's epistemic aims. So if one believes that the ultimate epistemic aim is believing truths and disbelieving falsehoods, then a justificatory norm is warranted insofar as it promotes belief in truths and disbelief in falsehoods. In the context of science, an epistemic consequentialist would say that a justificatory norm is warranted insofar as it promotes the discovery of true scientific findings and avoids the discovery of false scientific findings. This neat and simple story is not available to a nonconsequentialist philosophy of science. So what is the source of the normativity of justificatory norms according to a deontic philosophy of science?

Before proceeding, I want to clarify one point. A nonconsequentialist need not be unconcerned about consequences. Far from it. A deontic, nonconsequentialist philosopher of science can and ought to think that the outcome of science discovering a truth is good. In ethics, the term *good* is used to describe a valuable state, while the term *right* is used to describe appropriate action; the debate between consequentialist and deontological ethical theories can be construed as asking which is more fundamental, the good or the right, with consequentialists arguing that the good is more fundamental and deontologists arguing that the right is more fundamental. But of course, a deontologist can and should still say that good states are better than bad states, even if the right is more fundamental than the good. Same with deontic philosophy of science: one can reject epistemic consequentialism but still think, as one should, that discovering and believing truths is very important.

In our story about Sasha and Masha, we may ask: *Why* does Masha not endorse Sasha's method for assessing whether a number is prime? Suppose they were considering only numbers less than one million. About 8 percent of numbers below one million are prime. Thus, Sasha's method is 92 percent reliable. One might be tempted to respond by saying that 92 percent reliable is not sufficiently reliable and that is why Sasha's method should be rejected. Yet I do not believe that is right: such a response unreflectively maintains the sole concern about whether a method is truth-conducive to assess its epistemic merit. Suppose instead that Sasha and Masha are trying to determine whether a number is a "circular prime," which is a prime number with an additional property, namely, that any number generated by cyclically permuting its digits is also prime (so, for example, 17 is a circular prime because 71 is also prime; 3779 is a circular prime because 7793 is prime, 7937 is prime, and 9377 is prime). Now, suppose Sasha applies her same method to determine if a number is a circular prime. For numbers under one million, Sasha's method is now more than 99.99 percent reliable. Sasha and Masha then ask whether 745897 is a circular prime. Both say no. Sasha follows her usual method, while Masha reasons that a circular prime must contain only the digits 1, 3, 7, or 9, because any number with a digit 0, 2, 4, 6, or 8 would be, on some cycle of digits, divisible by 2, and any number with a digit 0 or 5 would be, on some cycle of digits, divisible by 5, and therefore 745897 cannot be a circular prime. So, again, Sasha and Masha share a weak consensus, but they do not share common knowledge because Masha does not endorse Sasha's justification for their shared conclusion about this number.

We are close to answering why Masha does not endorse Sasha's method, which will more generally indicate the source of normativity of justificatory norms for a deontic philosophy of science. The issue is not merely that Masha's justification is superior to Sasha's since scientists often must use second-best methods due to ethical, financial, or temporal constraints. Regarding truth-conduciveness, Masha's method is not obviously superior to Sasha's, nor is the issue that Sasha's method is error-prone about 0.01 percent of the time since most methods in science are less reliable than this. What, then, is the issue? Elgin (2017, 4) claims that "acceptable epistemic norms are norms that would emerge from the deliberations of suitably idealized epistemic agents." Justificatory norms (epistemic norms) are "norms of responsible epistemic agency" (2017, 91). Sasha's method would not be endorsed by the deliberation of idealized epistemic agents because her method does not reflect or promote epistemic agency, despite its truth-conduciveness. The rationality of Masha's method is attractive *not* primarily because it can successfully identify numbers that are not circular primes — unimportant for most people — but because it exemplifies good thinking more generally and manifests and conveys understanding in a way that Sasha's method does not. We might use Masha's method to teach students, not because we care about their ability to identify numbers that are not circular primes but because we want to cultivate rigor, rationality, and clarity of thought in their approach to the world more generally. Masha's method delivers not merely truth but also understanding, and many philosophers characterize understanding as a sort of intellectual "grasping," which is clearly agential (de Regt 2017; but see also Khalifa 2017). Sasha's method is heteronomous rather than autonomous, thoughtless rather than thoughtful, reckless rather than scrupulous, a quirk of habit rather than an expression of epistemic agency.

Discussing a person who neglects evidence, Clifford (1877, 294) claims, "The danger to society is not merely that it should believe wrong things, though that is great enough; but that it should become credulous, and lose the habit of testing things and inquiring into them; for then it must sink back into savagery." So, the norm of adjusting one's beliefs according to one's evidence is grounded not merely in the unreliability of not adhering to such a norm but also in the broader impact that not adhering to such a norm could have on epistemic agency and responsibility, and this is one problem with Sasha's method. Fundamentally, the problem with Sasha's method is that it is arational. A view that goes back as far as Aristotle holds that what distinguishes us from animals is our capacity for reason, which allows us to attain wisdom. Sasha's method does not use or develop this capacity for reason, and Masha's does. A group of idealized deliberators would not endorse Sasha's method, though they would endorse Masha's.

An appeal to the deliberations of a group of idealized agents is familiar in other branches of philosophy, such as ethics (Kant 1785) and political theory (Rawls 1971). We are asked to imagine that a group of idealized deliberators is informed about the relevant facts of a domain and tasked with articulating governing principles about that domain. Leibniz ([c. 1702–3] 1988, 56), for example, encouraged this approach: "Put yourself in the place of all, and suppose that they are well-informed and enlightened," and this will provide you with "the true point of view for judging what is just or not." So I call this the *Leibniz procedure*. In philosophy of science, Kitcher (2001) applied the Leibniz procedure to ask how the scientific research agenda should be determined, and Cartwright (2006) extended Kitcher's idea to argue that philosophers of science should focus on specific questions about the strengths and weaknesses of particular methods and how resulting evidence will be used, rightly suggesting that the Leibniz procedure can and should be sensitive to relevant contextual features. What fundamental considerations should this ideal group of deliberators appeal to when determining justificatory norms for science?

The deliberators in the Leibniz procedure are idealized in the sense that they are aware of relevant empirical facts about the domain and context they consider (so, for example, they know that rinsing test tubes three times eliminates more impurities than merely rinsing them twice), and they know that the domain they are deliberating about is obviously not ideal (to consider just one of many nonideal facts they are aware of, real scientists are not perfect and occasionally behave rather badly). The idealized deliberators are fully informed about philosophical considerations regarding the domain they consider (so, for example, when deliberating about what statistical tools are appropriate, they know that scientists should avoid the base-rate fallacy). They are ideal in their motives and are concerned strictly with epistemic properties (and not, for example, whether a putative justificatory norm will make people happier or wealthier).

We have already seen that one can reject epistemic consequentialism yet maintain that truth is important. Clearly, if the aim of science is common knowledge or any other factive notion, truth-conduciveness must be a desideratum for the deliberators in the Leibniz procedure. Yet truth-conduciveness is not the only desideratum that guides the idealized deliberators, as suggested by Masha's rejection of Sasha's method. The deliberators must ask themselves, as Kant had us ask ourselves, if a putative justificatory principle or practice is one that they would will to become a general maxim. Sasha's method is out; Masha's is in.

The problem with Sasha's method was that it was not an expression of epistemic agency, despite its high degree of reliability. Let us consider another example that motivates the importance of epistemic agency. Consider what would be gained if a mathematician developed an analytic proof of the four-color theorem. This theorem states that no more than four colors are needed to fill the spaces of a map such that no adjacent spaces contain the same color. The map can be composed of any pattern of crisscrossing lines and squiggles. The first proof of the theorem was a computer-assisted proof in 1976; since then, other such proofs have been developed, yet there still is no "fully human" analytic proof. We now know the theorem to be true and proven to be so—virtually all experts agree on this. Therefore, a new, analytic proof of the four-color theorem would not be truth-conducive since we already know that it is true. Yet we can nevertheless ask, If a mathematician were to construct such a proof, would that be good? I am not, of course, asking if the proof itself would be good (let us assume that it would be). I am also not asking if it is good that the mathematician has focused her time and talent on the four-color theorem rather than on some other problem. I am asking if it would be good if a new, fully human analytic proof of the theorem were constructed—not necessarily great or extremely good or even very good, but at least a little good, better than nothing. My inclination is to say yes. I would forgo at least one blueberry from the bowl of blueberries in front of me to bring that state about. Yet what kind of good would it be? A human, analytic proof of the four-color theorem would not add to our stock of truths, but it would manifest epistemic agency. Thus, the idealized deliberators in the Leibniz procedure have, in addition to their desideratum that justificatory norms be truth-conducive, a desideratum that justificatory norms manifest and promote epistemic agency.

The deliberators in the Leibniz procedure must also consider the role responsibilities of scientists. Roles, Hardimon (1994, 354) argues, are "institutionally defined clusters of rights and duties" that one is bound by in virtue of one's social role. Particular roles have particular role responsibilities. A father has a role responsibility to care for his children. A physician has a role responsibility that obliges her to be concerned with the promotion of the health of her patients. A friend has a role responsibility to listen to you when you need support.

Scientists have role responsibilities. Importantly, these responsibilities transcend the mere accumulation of truths. Perhaps one of the most obvious responsibilities is to not falsify data, though if epistemic consequentialism is true, then it is hard to see why falsifying data would be generally impermissible. Recall Sara, the scientist with the strange ability to intuit true physical laws. Suppose she is studying the Hubble constant, a measure of the expansion of the universe, and suppose the true value of the Hubble constant is 69 (km/s)/Mpc. Given Sara's special talent, she intuits the correct value, and she knows this value to be true. This involves an exercise of her epistemic agency. Her colleagues, however, do not accept this unusual method, as they cannot understand how it could be reliable. So Sara asks one of her PhD students to gather new evidence, which suggests that the value of the Hubble constant is 74 (km/s)/Mpc. Sara knows this estimate is incorrect, and she wants her colleagues to believe the truth, so (again, exercising agency) she manipulates some of the data to get it to indicate the true value. Sara's action here violates a role responsibility even though her action is truth-conducive. Thus the ideal deliberators in the Leibniz procedure must be concerned with both epistemic agency and epistemic responsibility when determining the justificatory norms for science.

Respect for each other as agents grounds the imperative for common knowledge. In a widely cited quote, Polanyi (1962) stated that "the authority of scientific opinion remains essentially mutual; it is established *between* scientists, not above them." Science, claimed Polanyi (1962), is a "chain of mutual appreciations" within which an individual scientist must "bear his equal share of responsibility." Interlocutors should be concerned with listening to the claims of others and the reasons offered for those claims and attempt to accommodate, critique, accept, or improve on those reasons because, as Rolin (2017, 2020) convincingly argues, scientists have epistemic responsibilities that arise out of an imperative for mutual respect.

So, justificatory norms in science, including domain-general principles and domain-specific practices, receive their normative status by being endorsed by a community of idealized deliberators in the Leibniz procedure. These deliberators admit justificatory norms by appealing to the desiderata of truth-conduciveness, epistemic agency, and epistemic responsibility.

The form that deliberation takes in the Leibniz procedure can be modeled dialogically, whereby a deliberator proposes a putative justificatory norm and another deliberator challenges its normative status. Such an approach has precedents. Longino's (1990, 71–72) emphasis on the importance of criticism is formulated dialogically, whereby one scientist criticizes the putative empirical or conceptual warrant for the claims of another scientist — her examples included Lewontin's reanalysis of the data that putatively supported Jensen's claim that IQ has a genetic basis, Gould's criticism of Barash's experiments that seemed to suggest that male mountain bluebirds punish their female mates for adultery, Einstein's rejection of the probabilities and discontinuities of quantum physics, and Millikan's rejection of Ehrenhaft's hypothesis of subelectrons. Similarly, Dutilh Novaes (2021, 49–54) recently offered a dialogical model of deduction that, she argues, can be thought of as a dialogue between a Prover, who wants to construct a proof of a deductive argument, and a Skeptic, who, as a stubborn interlocutor, continues to ask why a deductive argument is valid and who must be persuaded of its validity. In Longino's model, the dialogue occurs between actual scientists who argue about first-order features of the world—Lewontin and Jensen arguing about whether IQ has a genetic basis, for example. In Dutilh Novaes's model of deductive arguments, the dialogue is between interlocutors who argue about whether a particular deductive argument is valid and not about, say, whether a particular logical rule such as modus ponens is truth-preserving (the Skeptic grants that it is). While my model of common knowledge involves a first-order critical dialogue between actual scientists, as in Longino's model, who attempt to come to a consensus about a scientific claim and about its justification, in the Leibniz procedure the idealized deliberators must also come to agree about the normative status of putative justificatory norms, and this requires a second-order critical dialogue between the idealized deliberators about the justificatory norms themselves.

As noted in the introduction, some putative domain-general justificatory principles could include very basic norms like "reason according to the dictates of logic," "submit your work to peer review," and "falsifiable theories are better than unfalsifiable theories." Some putative domain-specific justificatory practices could include norms like "rinse test tubes three times with distilled water to remove impurities," "conceal the allocation of subjects in experimental groups so that unconscious biases of investigators cannot influence their interpretation of the outcomes," and "in high-energy physics report only results that surpass the five-sigma level of statistical significance." A set of justificatory norms in a particular domain can be thought of as an "institution" (Mantzavinos 2024)—as the rules of the domain—that provides solutions to problems of social coordination and dispute resolution. New justificatory norms can emerge in science as a result of novel problems, novel technologies, and novel conceptual or methodological innovations (in chapter 2, I describe the recent emergence of a novel justificatory norm). According to Mantzavinos, the rules constituting the institution of science are just the conventions adopted by actual scientific communities; my account of justificatory norms based on the Leibniz procedure is more idealized, as the justificatory norms are not merely the actual conventions of scientific disciplines but rather are a result of idealized deliberation grounded on desiderata relevant to the constitutive aim of science. Some conventional rules of actual science may not get through the Leibniz procedure, and, conversely, some deliverances of the Leibniz procedure may not yet be conventional rules.

Assessing the genuine normative status of a justificatory principle or practice may not be straightforward. While all deliberators in the Leibniz procedure will admit the fundamental requirements of logic, the rules of algebra, and the importance of avoiding biases and experimental confounds, the normative status of many putative justificatory principles and practices is not so obvious, which explains the vociferous debates that real scientific communities (and philosophers of science) engage in about many putative justificatory norms. For example, the platitude that simpler theories are better than complex theories has required a diligent and sophisticated philosophical investigation by Sober over decades to uncover precisely under what conditions the platitude indeed has normative status (see his 2015, for example). Similarly, the general applicability of Popper's falsifiability principle has long been challenged by philosophers. Moreover, just as in the ethical domain, the justificatory norms delivered by the Leibniz procedure will vary in their epistemic importance, and norms may trade off against each other. Therefore, in real scientific practice, how the satisfaction and violation of the many relevant norms ought to be balanced is not a straightforward matter.

Some justificatory practices can be evaluated empirically—for example, rinsing test tubes three times (rather than two times) with distilled water (rather than regular water) can be directly evaluated on empirical grounds for the practice's capacity to eliminate impurities. But the status of other justificatory norms may be much less straightforward to assess, in part because a putative justificatory norm's impact on truth-seeking is often not directly obvious, as illustrated by the debates about whether randomization is important for clinical trials (see, for example, Worrall [2002], who argues no, and Larroulet Philippi [2022], who argues, rightly in my view, yes). Indeed, assessing the normative status of some putative justificatory norms can require sophisticated and technical philosophical machinery, illustrated by an excellent article by Jäntgen (2023) on outcome measures.

Moreover, assessing the truth-conduciveness of empirical methods or other features of science requires knowing the reference standard—truth—yet in routine empirical contexts, that is impossible since the whole point of using the methods is to ascertain what the truth is (for this reason, computer simulations have recently begun to be used to evaluate scientific methods because, in these virtual worlds, investigators can stipulate the truth of a hypothesis and then vary features of a method and assess how close the method gets one to the truth; see, for example, Zollman [2007]; Romero [2017]; Tabatabaei Ghomi and Stegenga [2022, 2023]). Finally, some justificatory norms may themselves be philosophical in nature—examples could include "hypotheses should be consistent with well-established background theories," "proposed explanatory causal mechanisms should be plausible," and "measured physical parameters have single values." Justificatory norms of this sort are not amenable to empirical assessment but rather require philosophical evaluation.

That empirical evaluation of justificatory norms will not generally be sufficient further follows from the fact that the deliberators in the Leibniz procedure must ask not only whether a principle or practice is truth-conducive but also whether it manifests and promotes epistemic agency and responsibility, which is a philosophical question just as much as an empirical one. So, our community of idealized deliberators does not have an easy task. But this is the regulative ideal. The constitutive aim of science is common knowledge, and this entails that scientists must publicly articulate justificatory reasons for their claims, and the scientific community must critically engage with the justificatory status of those reasons, the normative status of which goes beyond truth-conduciveness. The task of the deliberators in the Leibniz procedure is, fundamentally, philosophical.

We can conceive of the resulting picture as a three-story epistemological structure. The ground floor is the realm of first-order claims about the world made by scientists and justified to a greater or lesser degree according to the extent to which the relevant justificatory norms are satisfied (for example, Neptune has fourteen moons). The next floor up is the realm of the Leibniz procedure, in which second-order claims about the status of the justificatory norms are assessed by the idealized deliberators (for example, to test a new drug, one must perform a randomized trial). The top floor is the realm of the desiderata that guide the Leibniz procedure (for example, truth-conduciveness).

It would be good to know who lives on each floor. Determining the residents on the ground floor is based on empirical and theoretical work by scientists. Determining the residents on the second floor can be an empirical or philosophical matter. Determining the residents on the top floor is a purely philosophical matter: I argued that the residents on the third floor include truth-conduciveness and manifestation and promotion of epistemic agency and responsibility. Because it is a philosophical matter to determine the residents on the top floor, the question of who lives there is liable to remain unsettled by the arguments I have offered. My arguments for including epistemic agency and responsibility depended on cases like those of Sara and Sasha, and I anticipate that a variety of intuitions will be elicited by such cases. Those who remain unconvinced by the examples of Sara and Sasha may think that the only resident on the third floor is the desideratum of truth-conduciveness, while others may agree that the third floor includes the desiderata of manifestation and promotion of epistemic agency and responsibility. If I am correct that the constitutive aim of science is common knowledge, then that also supports the thought that the third-floor residents include the manifestation and promotion of epistemic agency and responsibility since common knowledge requires a mutual understanding and acceptance of the justification of a scientific finding, and such mutual understanding itself involves the exercise of epistemic agency.

The account of common knowledge I have defended thus far suggests that Lipton's (2004, 1269) stance was incomplete in two important respects when he wrote, "What is required for knowledge is the de facto reliability of the methods we deploy, not our understanding of how those methods work or some demonstration of their reliability." The first problem with this claim is that the mere reliability of a method is insufficient to ground common knowledge because, in addition to a method's de facto reliability, scientists must be able to articulate the justificatory basis for a claim, and that requires understanding how the method that generated that justificatory basis works. The second problem is that even if a method is reliable, as Sasha's method was, and even if a scientist understands and can demonstrate the method's reliability, as we did on Sasha's behalf, that is insufficient to justify common knowledge because Sasha's method would not be endorsed by the Leibniz procedure. Many examples from the history of science show that mere de facto reliability is insufficient. When Galileo used his novel telescope to observe the moons of Jupiter, that was far from sufficient to convince his critics, even though we now know that Galileo's method was reliable. This is not to make the familiar point that scientific evidence is often insufficient to persuade stubborn dogmatists. The point is that putative claims to common knowledge are incomplete until their justification can be rendered acceptable by critical interlocutors. Galileo accepted this challenge and proceeded to articulate his justification. As a richer understanding of optics and the technology of telescopes developed, and as Galileo demonstrated the reliability of telescopes in familiar terrestrial contexts, the claim that Jupiter has moons became a secure contribution to common knowledge.

Similarly, Laudan's (1990, 318) stance was incomplete when he wrote, "Epistemic rationality, no less than any other sort of rationality, is a matter of integrating ends and means." Laudan was arguing that the normative status of a justificatory principle or practice is a function only of its conduciveness for one's epistemic aim. Laudan's "normative naturalism" encouraged us to look to history to determine whether a putative justificatory norm helps attain epistemic aims. For example, discussing the putative methodological rule that advises against ad hoc modifications of theories to resolve empirical disconfirmation, Laudan (1996, 137) complained that existing philosophical arguments involving mere pumping of intuitions amounted to "armchair bickering." Instead, he argued, we should examine history to see if scientists who upheld the methodological rule actually discovered truths more often than scientists who did not (see Knowles 2002 for discussion). We have seen reasons to doubt that epistemic rationality merely amounts to integrating ends and means. Moreover, Laudan claimed that aim-conduciveness is a question that requires only empirical investigation, which, as I argued above, is not the case.

Similarly, Kitcher's (2001, 123) stance was incomplete when he wrote that science should be concerned with identifying methods that are "maximally efficient." We saw above that Kitcher, developing his influential notion of "well-ordered science," also used the device of an idealized group of deliberators assessing scientific inquiry. The primary concern of the deliberators in Kitcher's well-ordered science was to determine the scientific research agenda, and he also had them determine moral constraints on science. The moral constraints of concern were ethical constraints on scientific methods, such as the protection of animal welfare and the rights of human subjects. As long as those moral constraints are satisfied, suggests Kitcher, the only concern of the deliberators when choosing among strategies for scientific investigation is to pick those that are most efficient. Yet I have argued that there is more to the normative status of justificatory norms than mere truth-conducive efficiency since, in addition to truth-conduciveness, justificatory norms should manifest and promote epistemic agency and responsibility.

A dogma in philosophy of science today is that there is no unique, unitary scientific method. Virtually every philosopher writing about the general nature of science in recent generations has promulgated this dogma. Mantzavinos (2024) offers an insightful response to this dogma by arguing that there is in fact a scientific method that is both unique to science and unitary in structure. The scientific method is constituted by what he calls the institutions of science. In my terms and with a suitable dose of idealization: The scientific method is just the set of justificatory norms delivered by the Leibniz procedure. That is an abstract, type-level characterization of the scientific method and, as we have seen, particular domains and contexts require distinct justificatory norms, which can be given a token-level characterization. Saying there is no such thing as "the scientific method" because there are many methods variably relevant to diverse contexts is like saying there is no such thing as money because there exist liras, rupees, pounds, and bitcoins that are variably relevant to diverse contexts (the analogy is thanks to Mantzavinos). This is consistent with pluralism about scientific practice, which is so popular among philosophers today (see, for example, Ruphy 2016 and Massimi 2022a), because clearly, multiple kinds of justificatory norms could be relevant to a particular research question. To repeat, following Mantzavinos's insight, I am tempted to say that the scientific method is just the deliverance of the Leibniz procedure.

The conditions for achieving common knowledge are very demanding. Some might think they are too demanding. One might think, with Lipton and others, that the only important condition for scientific knowledge is that claims are warranted by reliable methods. Yet the very demanding conditions for common knowledge are a virtue. We have seen that striving to satisfy these conditions creates many benefits for science. In any case, great scientists do not shy away from challenging demands—in Marie Curie's words: "I was taught that the way of progress was neither swift nor easy." The conditions for common knowledge are not so demanding that they cannot be met since the history of science is full of successful cases in which common knowledge was achieved: The double-helical structure of DNA, the dense nuclear core of atoms, and continental drift are three examples from the twentieth century. And once the conditions have been met, the resulting claim to knowledge is secure—successful cases of common knowledge are "timeless truths," a point I argue for in chapter 8.

The Leibniz procedure is not a philosopher's fanciful fiction. The critical exchanges about justificatory norms that it prescribes occur to some degree in science at its best, though of course those exchanges are between actual scientists, not a community of idealized deliberators. The advantage of adding this dose of idealization to the description of the Leibniz procedure is that it requires that the deliberators actively seek out and respond to criticism about justificatory norms. To put this in more down to earth terms, we can imagine a real scientific community that cultivates the sorts of critical practices that Longino (1990) promoted and actively seeks critical dialogue with scholars external to the community who are experts about justificatory norms themselves, such as philosophers of science. This occurs to some degree in science, as when the evidence-based medicine community attends to critical work by philosophers of medicine, or when open-minded physicists engage with philosophers of physics. As we will see in the chapters to come, the Leibniz procedure provides a foundation for science to achieve objectivity, value-free science, and timeless truths — ultimately, common knowledge, the constitutive aim of science — and provides guidance for real scientific communities to pursue that aim.

1.5 Objections

Here I respond to possible objections to my account of common knowledge, with the intention of adding some nuance to the notion.

Uninformative Common Knowledge

We saw in §1.3 the example of Aisha and Bheem, who reason from their respective justificatory norms to incompatible conclusions C_1 and C_2. A variety of possible strategies are open to them, including the mutual inspection and refinement of their justificatory norms, the gathering of further evidence that lends support to one of C_1 or C_2, or the articulation of a more coarse-grained conclusion C_3, which is entailed, perhaps implicitly, by both C_1 and C_2. For example, suppose Aisha and Bheem are climate scientists who have reviewed all the available evidence. Aisha concludes that by the year 2100, the global climate will increase by 2.4 degrees Celsius (C_1), and Bheem concludes that the global climate will increase by 2.8 degrees Celsius (C_2). They cannot agree simply by taking the mean of 2.6 degrees Celsius because this is outside their estimated ranges of error. To achieve common knowledge given their current resources, they could consider instead C_3: By the year 2100, the global climate will increase by at least 2 degrees Celsius. Given their justificatory norms, they both assent to C_3 and agree that their respective justificatory norms justify C_3. So Aisha and Bheem have common knowledge about C_3.

But this hedging approach to common knowledge raises a worry. Why stop at C_3? They could also consider C_4, which says that by the year 2100, the global climate will increase by at least 0 degrees Celsius, or they could consider C_5, which says that sometime in the future, the global climate will increase in temperature, decrease in temperature, or remain unchanged (Dethier 2024). The problem with C_5, of course, is that it is totally uninformative, and C_4 is nearly as uninformative, particularly for developing policies to mitigate the impact of climate change.

One way to achieve common knowledge is to develop a consensus about uninformative, vacuous propositions like "stuff existed yesterday," "stuff existed two days ago," "stuff existed three days ago," and so on. With a trivial amount of creativity, one could formulate a vast number of propositions about which there would be a strong consensus, all of which would be uninformative. Moreover, the hedging approach to common knowledge gives up some of the benefits of common knowledge, particularly the impetus to scrutinize and improve on justificatory norms. This is all antithetical to science.

To address this concern, we can turn to Levi's (1967, 113) claim that a fundamental cognitive goal is "relief from agnosticism," which one achieves by attaining true beliefs. Yet, as Levi (113) recognized, some true beliefs do not provide relief from agnosticism because they are uninformative, as we saw with C_5 and the series of propositions about stuff existing in the past. To achieve relief from agnosticism, the content of the proposition meant to provide the relief must be informative: "Highly informative conclusions are desirable because they settle questions" (and not, contrary to Popper, simply because they raise new questions or because they can be severely tested). Because informative propositions are more likely to be false than uninformative propositions, we must take epistemic risks when seeking relief from agnosticism (119).

One could live like a monk in a cave, refusing to believe or assert anything. This would avoid epistemic risks, but then one's status as a troglodyte would effectively exclude them from being a responsible member of a scientific community. As I have argued, the aim of science is common knowledge, and scientists must actively try to contribute to Leibniz's public treasury; the demand that scientific claims be informative constrains the injudicious use of a hedging strategy to achieve common knowledge. That is, a reasonable constraint on the hedging strategy to achieve common knowledge is that it should be used sparingly, leaving a hedged conclusion as informative as possible despite the hedging. That would warrant the move to C_3 in the above example but not to C_4 or C_5. This constraint is not merely an ad hoc patch but rather is central to the notion of common knowledge. Kitcher (2001) convincingly argued that idealized deliberators setting a research agenda for science would promote the pursuit of "significant" truths, and significance itself can be explicated as informative for inference or for action (in chapter 4, I develop the notion of informativeness). Excessively hedging scientific conclusions saps them of their significance. While hedging claims is often the responsible thing to do as scientists, excessively hedging a claim such that it becomes uninformative thwarts one's epistemic responsibility and ability to contribute to the public treasury of common knowledge.

Dissent, Productive and Stubborn

Dissent can be productive. Feyerabend argued for what he called a principle of proliferation, which held that scientists should invent and elaborate theories that are inconsistent with consensus views (Shaw 2020). Popper (1994, x) seemed to believe something similar: "The growth of knowledge depends entirely upon disagreement." Given the potential for dissent to contribute to scientific achievements, a point insightfully argued by Solomon (2001) and going back as far as Mill (1859), one might think that my emphasis on consensus in common knowledge is misguided. Conversely, some forms of dissent are stubborn: Dissenters resist assent to a consensus on unreasonable grounds, perhaps for personal, political, or financial reasons (Oreskes and Conway 2010). Dutilh Novaes (2021, 58) rightly notes that even in response to a deductive argument, there can be dissenters, or "cantankerous opponents" to use Aristotle's term, who refuse to grant argumentative moves even when they are genuinely compelling. Stubborn dissenters, whether their motivations are genuinely epistemic or motivated by nonepistemic reasons, can take cover as productive dissenters—a scientist who doubts the harmful effects of smoking can cast themselves as another Galileo being unjustly persecuted. It follows that consent can be blocked for nonepistemic reasons by stubborn dissenters.

One might think that dissent causes two problems for my account of common knowledge. First, by its emphasis on consensus, common knowledge might overlook the benefits of productive dissent. Second, by its requirement of strong consensus, common knowledge might be inappropriately blocked by stubborn dissent, and thus, particularly for policy-relevant science, common knowledge is too demanding.

My response to the first concern is that dissent should not be seen as an end in itself but rather as a means for exposing problems in existing scientific work, the resolution of which would amount to scientific progress and could result in consensus. Transient dissent should be seen as a means to the end of improved justification, and that in turn can have a long-run consensus-forming effect. For example, dissent about a sun-centered (heliocentric) model of the solar system was one motivation for Galileo to develop the telescope, which ultimately improved the justificatory status of the sun-centered model. While achieving strong consensus about a sun-centered model of the solar system took many decades, once achieved, it became wildly implausible to dispute; today, one must have extreme and implausible views to deny a sun-centered model of the solar system. Dissent about a particular scientific claim is no longer productive once that claim has the status of common knowledge.

My response to the second concern is that while common knowledge is the constitutive aim of science, not achieving it for a particular hypothesis is not a failure. As I argued in the introduction, science should be judged on deontic grounds rather than consequentialist grounds. Not reaching common knowledge because of stubborn dissenters is not grounds for negative judgment on the science in question, and such science can, despite the presence of dissent, nevertheless be the basis for policy. Moreover, as noted earlier, the strong consensus required by common knowledge need not be perfect unanimity, and one can expect a minority of dissenters in a large enough scientific community, particularly for science as it plays out in contemporary research. Yet for established scientific knowledge, such as the sun-centered model of the solar system, dissent is best seen as a problem with the dissenter rather than a failure to achieve common knowledge.

Rules Are Meant to Be Broken

There is a pronounced suspicion in philosophy of science about general or universal rules or norms about scientific methodology. Feyerabend (1975) argues that "anything goes"; that is, science should not be norm-bound. Norton (2021) argues that there is no universal inductive rule for science but that all rules are local, determined by the particular material facts in a domain. Cartwright (2006) argues that justificatory norms should be informed by how evidence is intended to be used—for example, a randomized trial can tell us if an intervention is effective in a controlled setting but is often a poor guide to informing us whether that intervention will be effective in an uncontrolled, real-world setting. Rules are meant to be broken.

Yet I do not think these considerations are an obstacle for a deontic philosophy of science. Feyerabend's anarchism was ambiguous regarding its positive recommendation—some understood his thesis as the modest view that scientific standards change over time, some saw it as a form of methodological pluralism, and some interpreted it as meaning that all methodological rules are useless (for discussion, see Russell 1983). The modest interpretations of Feyerabend's anarchism can easily be accommodated by the deontic philosophy of science on offer here, while the radical interpretations can be rejected as implausible. The context sensitivity urged by Norton and Cartwright is compelling and can be accommodated by a deontic philosophy of science. There is no principled barrier for the deliberators in the Leibniz procedure to consider context-specific details. Insofar as any philosophical argument about scientific methods is persuasive, the deliberators will consider that argument when formulating justificatory norms. So, for example, nothing stops the deliberators from attending to Cartwright's lesson that when formulating methodological principles, we should account for how evidence will be used in policy settings.

Scientists Are Not Ideal

Regarding the normative status of justificatory norms, I asked what a group of idealized deliberators in the Leibniz procedure would say about any putative justificatory norm. But scientists are not ideal.

Rawls (1971) used a technique like the Leibniz procedure. He imagined a group of idealized deliberators tasked with articulating rules that instantiate a theory of justice. Rawls developed an "ideal theory" for justice, which assumed "strict compliance" by people to whatever rules are formulated in his ideal theory. Yet Appiah (2017) argues that rules formulated according to the assumption of strict compliance might be less just in actual societies in which compliance with rules is partial. The idealized deliberators should, Appiah states, consider the degree to which compliance with rules can be expected. Mills (2005) goes further, cautioning that ideal theory can be masked ideology.

While this concern about ideal theory probably applies more to ideal political theory than it does to ideal philosophy of science, it raises an important consideration for the Leibniz procedure. The concern about ideal theory challenges the completeness and relevance of the set of norms that would be delivered by the Leibniz procedure if full compliance (and other idealizations) is assumed. Yet our deliberators, though they are idealized, are not deliberating about ideal circumstances. Rather, they are informed about nonideal features of real science. Scientists are motivated by all sorts of things other than truth, including fame and money, and they are not perfectly adept, so they sometimes violate justificatory norms. For example, one plausible domain-general justificatory principle might be "always be honest and transparent about your methods and results."

Yet, regarding honesty, to stop there would be to engage in ideal theory. Valentini (2012) argues that for philosophical theory to be action-guiding in the real, nonideal world, philosophical theory should account for the fact that some agents will not follow governing norms. We know that scientists too often engage in practices like p-hacking (selectively analyzing data to get one's desired result without fully disclosing one's method), and — much more rarely but much more egregiously — some scientists engage in outright deception and fraud; such are the corrupting incentives in science. This suggests that the deliberators in the Leibniz procedure should articulate norms that, in addition to a general norm of honesty and transparency, provide exogenous encouragement or enforcement of honesty and transparency. The recent requirement in clinical research and psychology of preregistering one's experiments can be seen as precisely such an honesty-enforcement mechanism (I discuss this in chapter 2).

Utopian Aims

Laudan argued that (i) truth is a utopian aim and (ii) the pursuit of utopian aims is irrational; therefore, since science is rational, truth is not the aim of science. If this argument is sound, then the aim of science cannot be a type of knowledge because knowledge implies truth, and so the aim of science cannot be common knowledge.

By utopian aim, Laudan meant an aim that is impossible to achieve or impossible to know that we have achieved or are approaching it. Laudan defended (i) in both senses of utopian. Most famously, the pessimistic meta-induction was intended to show that science, a graveyard of dead theories, cannot attain truth. Many philosophers have since criticized the pessimistic meta-induction, rendering the defense of (i) on the first sense of utopian dubious (see chapter 8 for discussion and references).

Appealing to the other sense of utopian, Laudan (1984, 53) claimed, "If we cannot ascertain when a proposed goal state has been achieved and when it has not, then we cannot possibly embark on a rationally grounded set of actions to achieve or promote that goal." This was a popular sentiment among late twentieth-century philosophers. Rorty (1995, 39), for example, wrote that truth is not a goal because truth "is neither something we might realize we had reached, nor something to which we might get closer." Davidson (2000, 67) echoed this thought: "Truths do not come with a 'mark,' like the date in the corner of some photographs, which distinguishes them from falsehoods. The best we can do is test, experiment, keep an open mind.... Since it is neither a visible target, nor recognizable when achieved, there is no point in calling truth a goal." (See Chrisman [2022], especially chapter 5, for a discussion of these claims by Davidson and Rorty in the context of norms for individuals' beliefs.)

However, one can have a goal for which one can identify the means to promote it even if one cannot determine when it has been achieved. Suppose I want to lose enough weight so that I weigh less than seventy kilograms, yet I am stranded on a deserted island that is rich in creamy coconuts, and I do not have a scale. I can take steps to promote my goal— eat fewer coconuts!—even though I will not be able to tell when I have achieved my goal. So, premise (i) in Laudan's argument is shaky: Truth is not a utopian aim in either sense of utopian. Premise (ii) looks equally implausible. World peace may be utopian, but its pursuit is not irrational, and we can take concrete steps toward it. Both of Laudan's premises are unconvincing, and thus truth survives as a plausible aim of science. We saw in the introduction that physicist Lise Meitner, who contributed to the discovery of nuclear fission, claimed that science is "a battle for final truth," and I do not think it wise to contradict her.

Yet there is an important insight in the thought that truth is a utopian aim. Holding truth to be the aim of science, Laudan (1977, 127) argued pragmatically, is "not very helpful if our object is to explain how scientific theories are (or should be) evaluated." This is roughly for the reason sketched above: If we cannot ascertain in real time when truth has been achieved, then we cannot use the ascertainment of truth for real-time evaluation of science. Rorty (1995, 281) put the point nicely: "Assessment of truth and assessment of justification are, when the question is about what I should believe now, the same activity." For these reasons, truth is a *nirvana norm*. I briefly mentioned this notion in the introduction, and I return to it in chapters 4 and 5. A nirvana norm gives little guidance for action and little basis for evaluation. Even if truth or a special kind of truth such as common knowledge is the aim of science, stating that fact does not help much in evaluating first-order features of science. In chapter 2, I give an account of scientific evaluation, deontic evaluation, which avoids direct reference to the attainment of truth as a success condition for science, is action-guiding, and provides an epistemically accessible basis for evaluation.

1.6 Conclusion

Common knowledge is the constitutive aim of science, and it is achieved when there is a strong consensus about a scientific claim.

That common knowledge is the aim of science unifies a wide number of insights in philosophy of science. That common knowledge is the constitutive aim of science explains why Dellsén (2016), de Regt (2017), Elgin (2017), and others maintain that understanding is an important aim of science because a plausible requirement for understanding is that one be able to articulate some justification for one's object of understanding, and the public articulation of justification in common knowledge affords that. Moreover, understanding requires epistemic agency, so to maintain that understanding is important for science lends support to the thought that epistemic agency is a criterion for the normative status of justificatory norms (Elgin 2017). That common knowledge is the constitutive aim of science supports Gerken's (2022) view that scientific testimony must be justified and such justification must be publicly articulated. That common knowledge is the constitutive aim of science explains many scientific practices, particularly the publicity of putative justifications for a claim and the scrutiny and criticism of those putative justifications. That common knowledge is the constitutive aim of science lends support to Kusch's (2002) communitarian epistemology while avoiding the pitfall of holding justificatory norms to be merely social norms. That common knowledge is the constitutive aim of science explains why Longino's (1990) "critical empiricism" has been so influential.

In contrast, that common knowledge is the constitutive aim of science generates tension with some existing views in philosophy of science. That common knowledge is the constitutive aim of science contradicts Massimi's (2022a, 363) claim that "whatever the aims of science might be, they are likely to be localized and discipline-specific." That common knowledge is the constitutive aim of science suggests an alternative organizing principle to Kuhn's notion of paradigms since common knowledge is a domain-general aim, and, as I argue in chapter 8, contributions to common knowledge are not liable to be overturned in a revolutionary manner. That common knowledge is the constitutive aim of science offers a fresh version of the value-free ideal for science, as I argue in chapter 3, thereby resisting what has become a new dogma in philosophy of science.

Once a scientific finding becomes common knowledge, it becomes part of a second-order justificatory norm, namely, a principle that future scientific claims be consistent with established common knowledge. Violation of that norm is common in pseudosciences; for example, the underlying idea of homeopathy is that a substance infinitely diluted in water can have beneficial physiological effects via the putative "memory" of the water, yet that contradicts very basic features of physical chemistry that are now part of common knowledge. To be clear, this principle of consistency with common knowledge is not an overly conservative principle that would never permit the development of scientific theories that were inconsistent with existing knowledge; it is rather simply a desideratum on routine science.

The normative status of justificatory principles and practices is based not merely on truth-conduciveness but also on the manifestation and promotion of epistemic agency and responsibility, as I argued in §1.4. The deliberators in the Leibniz procedure appeal to those three desiderata to determine the set of genuine justificatory norms for science, thereby providing the foundation for the deontic philosophy of science proposed in the introduction, the implications of which I explore in the remainder of this book. The Leibniz procedure provides an especially sturdy foundation for a new approach to scientific evaluation and related notions such as objectivity, bias, and trust in science, which I explore in chapter 2.

CHAPTER TWO

Deontic Evaluation

2.1 Introduction

Some of the most important and long-standing tasks for philosophy of science have been to articulate principles to help people determine what to believe based on scientific findings, what to do based on scientific advice, and how to use science to structure aspects of society. These aims of philosophy of science are ancient and have accompanied modern science from its moment of birth. When Copernicus published his *De Revolutionibus Orbium Coelestium* (*On the Revolutions of the Celestial Spheres*) in 1543, Andreas Osiander, who was managing its publication, added a preface, unsigned and unauthorized, telling readers that the sun-centered model proposed in the book was merely a useful calculating device and not to be believed as true. At an abstract level, these tasks for philosophy of science are unified by their evaluative nature: One way or another, they assess which epistemic projects are reliable for guiding inference and action.

In the twentieth century, one of these tasks of philosophy of science came to be known as the problem of demarcation. Philosophers sought a principle of demarcation that could serve to distinguish genuinely scientific beliefs, claims, and practices from other sorts of beliefs, claims, and practices, such as those from religion, mysticism, or pseudoscience. This aim, most philosophers now believe, has not been achieved. The project of articulating a demarcation criterion has been a failure, according to many philosophers of science, since all general theories of demarcation face insurmountable problems. In 1983, Laudan published an essay titled "The Demise of the Demarcation Problem," and indeed, by the 1980s, the project of demarcation had seemingly died.

Yet the original impetus for that project is just as pressing today. We rely on scientific experts to guide us in many aspects of life, like what to eat and when to sleep, how to manage features of our mental lives such as anxiety or depression, and how to structure elements of society to mitigate the impact of climate change. Consider a recent and very impactful example: the respiratory coronavirus that caused the COVID-19 pandemic. It would be very good to know whether the virus escaped from a laboratory dedicated to gain-of-function research or if it came from animals, whether masks are effective at slowing the spread of respiratory viruses, what the long-term side effects of coronavirus vaccines are, whether epidemiological models are reliable guides to the future dynamics of a pandemic, whether closing schools and businesses and borders significantly slows the spread of the virus, and whether there are helpful pharmaceutical treatments for coronavirus infections. The science relevant to these questions is vast, complex, contradictory, and highly variable in quality. Though the demise of the demarcation project entails that we cannot simply sort some of this scientific work into a genuine-science bin and the rest into a junk-science bin, that is for the best, as the relevant science varies on too many dimensions for such a simple sorting. Yet it remains crucially important to have a basis for evaluating science.

The demarcation project was, clearly, a project of evaluation. Deeming an endeavor scientific amounted to saying that one should hold a positive attitude toward that endeavor and the claims asserted in that endeavor, whereas deeming an endeavor nonscientific amounted to saying that one should hold a less positive attitude or even a negative attitude toward that endeavor and its corresponding claims. When the grand evaluative project of demarcation died, philosophy of science did not, thankfully, give up on evaluative projects altogether. Instead, philosophy of science shifted focus to more local evaluative projects. Instead of seeking a single, domain-general evaluative principle to underwrite the demarcation of good science from bad, philosophers of science began to ask domain-specific questions, such as whether randomized trials are necessarily more reliable than non-randomized trials, whether idealized models can teach us about the world, whether experiments in psychology and medicine should be preregistered, whether particular kinds of statistical tests are appropriate, or how to best interpret various concepts in science such as gene, life, cause, and time.

In this chapter, I bring together the fundamental ambition of the demarcation project, namely, the articulation of domain-general principles with which one can evaluate science, with the local nature of the more recent evaluative projects in philosophy of science. In chapter 1, I articulated the fundamental basis of the normativity of justificatory norms, and in the introduction, I argued that our evaluative concepts for science should be based on justification. Demarcation has been the evaluative concept par excellence in philosophy of science. The deontic philosophy of science developed in the introduction and chapter 1 provides the foundation for the evaluative project pursued in this chapter.

The project of demarcating science from pseudoscience appears to involve a binary evaluation based on a single property of finished products—an evaluation of theories or statements based on the single property of "falsifiability," to take Popper's criterion as an example. That property—which asks whether there is a possible way to refute a theory or statement—is ungraded (though Popper was aware of the need for a graded approach to demarcation; see Bárdos and Tuboly, 2025). These three features of those early demarcation projects—product-oriented evaluation, evaluation based on a single feature, and ungraded evaluation—are damning shortcomings for the evaluation of an endeavor as multifarious and complex as science.

An evaluation of science can involve the assessment of processes rather than products, it can appeal to multiple evaluative principles rather than one grand principle of demarcation, and it can be graded. Consider an analogy. Suppose a professor must demarcate her students into two groups: those who pass her course and those who fail. One option for this professor would be to adopt a single method of assessment with a predetermined threshold for passing. That option has the benefit of simplicity, but that is also its drawback since it is unrealistic to assume that student performance is a unidimensional property that can be measured with a single mode of assessment. A better option for this professor would be to use multiple methods of assessment such as exam performance, quality of essays, participation, effort, and perhaps even second-order properties such as improvement over the duration of the course.

The evaluative project of demarcation had the aim of distinguishing science from some nonscience alternative. The domain distinguished from science by a demarcation principle could be completely unscientific, such as religion, or the distinguished domain could be pseudoscience or junk science; that is, it could have the look and feel of science yet be not quite scientific, such as psychoanalysis (one of Popper's favorite targets). As Laudan (1983, 118) put it in his obituary for the demarcation project: "A philosophical demarcation criterion must be an adequate explication of our ordinary ways of partitioning science from non-science and it must exhibit epistemically significant differences between science and non-science." Such an approach to the evaluation of science entails that one cannot use a demarcation criterion to evaluate domains that are deemed scientific by the demarcation criterion but have features that are epistemically problematic, such as an epidemiological model with parameters that have mediocre empirical warrant. Moreover, it is not obvious that there are ordinary ways of partitioning science from nonscience, though, of course, there are clear token instances on either side of any plausible partition (twentieth-century atomic physics on one side, tarot-reading psychics on the other). Finally, as Hansson (2021) writes, the demarcation project "is part of the larger task of determining which beliefs are epistemically warranted." Yet precisely because science involves pushing the boundaries of inquiry to its limits, there are many scientific theories for which we should have an epistemically cautious attitude, such as an estimate of the Hubble constant or a putative cause of the extinction of dinosaurs or any version of string theory, while there are many nonscientific domains for which we can be epistemically confident, such as my belief that there is beer in my fridge (a point emphasized by Laudan). So, a simplistically demarcated science is a poor guide to epistemically warranted belief.

Laudan believed the implication of these considerations was that the attempt to distinguish science from nonscience is misguided. The project of evaluation should be framed around questions like "What should we believe?" or "Based on this research, what should we do?" or "Should we trust this person when she claims x?" Those questions clearly apply to domains that are not scientific just as much as they apply to science. They are, of course, evaluative questions. They direct our attention to the reasons, scientific or otherwise, that provide putative support to a claim under question. In a scientific context, all the relevant domain-general justificatory principles and domain-specific justificatory practices identified in the Leibniz procedure (described in chapter 1) can and should be considered to answer such questions. Because belief itself is very plausibly graded, and because our attitude of trust is graded, our response to the above questions about belief and trust can be graded. This can be called *deontic evaluation*, given the centrality of justificatory norms for this evaluative project, and given that our evaluation of scientific statements in real time is never based on truth itself (for reasons given in the introduction and described further in chapter 4). I conceived of the intellectual project of my first book, *Medical Nihilism*, as "contextualised demarcation" (Stegenga 2018, 3), which is similar to deontic evaluation: Evaluation of science should be contextual and based on the extent to which relevant justificatory norms are satisfied. Deontic evaluation has many advantages over various versions of the traditional demarcation project.

In their epic study of the history of objectivity in science, Daston and Galison (2007) argue that the way scientific objectivity has been conceived has changed over time and the various accounts of scientific objectivity in different eras are influenced by one's epistemological fears. For example, in the nineteenth century, scientists became more concerned with the biasing influence of human perception and cognition, perhaps due to Kant's influence on epistemology. Therefore, scientists of the day emphasized "mechanical" objectivity by, for example, depicting objects with photographs rather than hand-drawn pictures. Something similar could be said about scientific evaluation in general. The specific shapes taken by programs of scientific evaluation reflect epistemological, methodological, and even political concerns. In the final decades of the nineteenth century and the first decades of the twentieth century, physics, biology, chemistry, and medicine were in a golden age. Ingenious experiments and theories taught us about fundamentally important aspects of the world, including the nature of matter at atomic scales, the germ theory of disease, and the evolutionary origin of complex life. At the same time, there was little of the scandalous corruption that pervades much of science today—writing in 1942, sociologist Robert Merton (1973, 276) observed a "virtual absence" of fraud in science. Of course, some areas of science in this era, such as eugenics and phrenology, were based on sloppy empirical methods and pernicious and implausible theories, but atomic physics, physiological medicine, and other areas of science were arguably at a zenith. During those same decades, though, a specter of ideas haunted the world: Nationalism and communism were implicated in the horrors of these decades, from the trenches of World War I, to the Holodomor famine imposed by Soviet Russia on Ukraine, to the rise of the German National Socialist Party and the resulting Holocaust. In the midst of this, Popper was deeply worried about the political influence of Marxism, and, since Marx himself claimed that his criticism of capitalism in *Das Kapital* was scientific, Popper used his demarcation principle of falsifiability to criticize it. In short, the general merits of science in the early decades of the twentieth century, combined with the horrors resulting from ideologies, provided compelling reasons to articulate a simple evaluative principle that could be used to resist such ideologies.

Today, the threat of ideologies is as great as ever. Today, though, we must also be concerned about science itself. Science today is shot through with dubious theorizing and sloppy methods that generate unreliable evidence, causing what psychologists call a "replication crisis" and leading prominent medical authorities to claim that anything written in medical journals cannot be trusted (Frances 2015). So-called paper mills sell authorship of scientific articles in a wide range of disciplines based on fabricated or plagiarized data on an industrial scale. During the writing of this book, the president of Stanford University resigned his post after being accused of manipulating evidence, thereby becoming a prominent symbol of the sorry state of so much of science today.

Yet, science remains our best means of learning about the world, and thus it is as important now as ever before to take a nuanced approach to evaluating science. In §2.2, I describe several of the most prominent approaches to demarcation and some of the major problems with those approaches. In §2.3, I present my own approach, which holds that we should evaluate science based on the extent to which the domain-general and the domain-specific justificatory norms delivered by the Leibniz procedure (described in chapter 1) are satisfied. In §2.4 and §2.5, I argue that this approach to scientific evaluation can also offer compelling accounts of scientific objectivity, bias in science, and conditions for trust in science. New justificatory norms arise at particular moments in science, and so in §2.6 (a section that is somewhat parenthetical to the rest of the chapter), I describe an instance of the emergence of a new norm in recent science, particularly in psychology and medicine, namely, that one should preregister experiments. I close in §2.7 with a brief recap of my positive proposal for replacing the single-criterion demarcation approach to evaluating science, developed by Popper and others, with an approach to evaluating science that assesses the extent to which relevant justificatory norms have been satisfied.

2.2 Demarcation Projects

In this section, I describe some of the most prominent accounts of demarcation. This survey can be skipped by readers who are familiar with the relevant material.

The logical positivists were concerned with demarcating meaningful statements from meaningless statements. According to the positivists, the meaning of a statement is its method of verification, and statements that cannot be verified are meaningless. Their view was that there are two classes of verifiable statements: logical claims and empirical claims. So, putting aside logical claims, the ultimate basis of meaning is empirical experience. Thus many religious, moral, and metaphysical claims are meaningless according to this criterion (see Richardson [2023] for discussion). For example, the claim "the absolute spirit is perfect" is not verifiable and thus is meaningless on the positivists' account, but the claim "metals expand when heated" is verifiable and thus is meaningful. Notice, however, that the disjunction of these two claims is also verifiable because if the latter can be verified, then the disjunction can be verified too (Godfrey-Smith 2003, 31). That sort of logical problem plagued the positivists' attempts to formulate a criterion of meaningfulness and, hence, of demarcation.

Popper argued that it is too easy to find putative verification of theories or statements about the world as long as the statements are ambiguous enough or one permits ad hoc adjustments to the content of one's theory. Impressed by Hume's problem of induction, Popper claimed that no theory could ever receive inductive confirmation. So, famously, Popper (1959) offered his principle of falsifiability. A genuinely scientific theory takes bold risks and can possibly be falsified by evidence. Good scientific practice focuses on trying to falsify a theory rather than looking for evidence to verify it. If a theory survives rigorous attempts at falsification, it can be tentatively included among accepted scientific theories without claiming that it is thereby confirmed or likely to be true. Einstein's prediction that gravity bends the path of light to a degree different from that predicted by classical physics was a shining example of a risky prediction. In contrast, argued Popper, the theoretical apparatus of psychoanalysis was sufficiently ambiguous and flexible to accommodate any evidence. Einstein's prediction was falsifiable, but psychoanalysis was not. Popper's ambition was to demarcate genuine science from pseudoscience, which, according to Popper, included Freudian psychoanalysis and Marxist theory of history.

Critics of Popper's criterion of falsifiability argued that evidence that seems to falsify a theory might in fact arise due to a faulty auxiliary hypothesis rather than the theory being false (a point Popper recognized). Some critics also claimed that there are scientific disciplines that are uncontroversially not pseudosciences and yet involve theories that can appear to be unfalsifiable, such as evolutionary theory (for example, the absence of "missing link" fossils does not cause biologists to abandon the theory of evolution but rather prompts them to offer explanations for why missing link fossils are missing—as Darwin did in *On the Origin of Species*—and motivates them to hunt further for missing link fossils, of which they have now found many). Another criticism was based on Popper's aversion to confirmation. On Popper's account, theories could not be confirmed or accrue any degree of incremental confirmation, leading to the strange result that the theory that water is H_2O is just as unconfirmed as some exotic novel theory introduced last week in high-energy physics. Yet another problem with Popper's falsificationism was its strict opposition to scientists' attempts to verify their theories; at least sometimes, such behavior can lead to great discoveries. For example, when Marie Curie observed that uranium ore was more radioactive than uranium itself, she hypothesized that uranium ore contained another much more radioactive element, which prompted in Curie "a passionate desire to verify this hypothesis as rapidly as possible" (quoted in Reid 1974, 83). That desire to verify her hypothesis led her to discover radium and polonium.

Lakatos (1978) proposed a significant revision to falsificationism. His account, "sophisticated (methodological) falsificationism," held that demarcation should not be about individual theories, as in Popper's account, but rather should be about an entire research program. Lakatos's primary distinction was between what he called "progressive research programmes," which involve a sequence of theories making more and more novel predictions that are ultimately confirmed, and "degenerating research programmes," which involve a sequence of theories that are developed and modified only to accommodate empirical anomalies. Twentieth-century genetics is a good example of a progressive research program, while medieval Ptolemaic astronomy is a good example of a degenerating research program. While this is an insightful development in thinking about the nature of science, a key problem with Lakatos's account is the weight it places on novel predictions. The view known as *predictivism* holds that hypotheses that are designed merely to accommodate existing evidence receive less confirmation than those that are developed before the observation of that evidence and that predict that novel evidence (Barnes 2008). The recent philosophical literature on predictivism has more or less settled that the prediction of novel evidence sometimes provides more confirmation to a hypothesis than the mere accommodation of that evidence, but not always (see, for example, Frisch [2015]; Hitzig and Stegenga [2020]). Lakatos's criterion depends on a single principle that is itself not generally true. Moreover, there are other important epistemic considerations besides the prediction of novel evidence, such as consistency with existing theories, breadth of applicability, and indeed, falsifiability, so Lakatos's focus on a single epistemic consideration was excessively narrow.

One of the most recent attempts to offer a single domain-general demarcation criterion comes from Hoyningen-Huene (2013), who claims that scientific knowledge is distinguished from other kinds of knowledge based on its systematicity. Critics responded by arguing that systematicity does not distinguish science from other kinds of endeavors because nonscientific endeavors can also have some systematicity, such as stamp collecting (see, for example, Thalos 2015).

The failure of single-criterion principles of demarcation has led some philosophers to develop multicriteria approaches, a positive development that (in a sense) my approach builds on. An early and well-known example of this approach is from sociologist Robert Merton ([1942] 1973), who argued that science has four institutional norms: universalism (acceptance or rejection of hypotheses should not be based on a particular individual's idiosyncrasies), communism (science is a communal enterprise, and therefore its products belong to the community of science), disinterestedness (scientific institutions are designed to promote the genuine aims of science, not personal or ideological interests), and organized skepticism (science involves structured practices of criticism). A significant development in thinking about scientific evaluation came from Longino (1990), who noted that the various practices of criticism in science help science to attain some degree of objectivity. Because criticism can target various aspects of science, Longino's account is effectively a multicriteria theory of scientific evaluation.

More recently, Resnik and Elliott (2023) offered a list of what they call "scientific norms," which, like Merton's list, comprises a mix of both epistemic norms, such as "rigor" and "evidentiary support," and ethical norms, such as "fair sharing of credit" and "respect." This approach is compelling insofar as it recognizes that evaluating science should be based on multiple properties rather than a necessary or sufficient condition or set of conditions and because it recognizes that evaluating science need not result in a binary judgment but can and should be graded. However, by mixing both ethical and epistemic properties in their list of scientific norms, Resnik and Elliott's approach appears haphazard. For instance, when giving examples of different kinds of norms, their categories include "policies for assigning authorship on publications" and "policies pertaining to hiring, tenure, and promotion" alongside "rules or conventions for designing experiments" (see 2023, 276, table 2). While the former sorts of considerations are probably important for science to perform well and fairly, they are irrelevant for the epistemic evaluation of science. Moreover, Resnik and Elliott do not explain where their various policies and norms get their normative status, so their list of rules resembles a collection of comforting truisms. Yet it is a step forward in thinking about scientific evaluation precisely because their approach encourages considering multiple relevant criteria and delivering a graded assessment.

In the following section, I provide a general framework for scientific evaluation that combines the ambition of earlier demarcation projects to provide a domain-general and unified schema of judgment while acknowledging that there are multiple domain-general and domain-specific properties that are relevant to the epistemic evaluation of science.

2.3 Judging Science

The demarcation project was always about whether one should have a positive evaluative attitude toward a particular statement about the world or a general domain of inquiry. After this preamble and the groundwork laid in the introduction and chapter 1, there is little left to reveal about my proposal to rethink the demarcation project. My account of scientific evaluation combines the best of the earlier demarcation projects and the more recent multicriteria accounts while avoiding their shortcomings. The evaluation of science should be based on the domain-general and domain-specific justificatory norms that are delivered by the Leibniz procedure described in chapter 1. Whatever those justificatory norms turn out to be, they form the basis for the judgment of science. For some unit of scientific work, we simply must ask: To what extent have the relevant justificatory norms been met? For example, if the hypothesis under investigation is about the benefits of a new pharmaceutical, then, in addition to whatever domain-general norms apply, a putative domain-specific norm is to test pharmaceuticals with randomized trials, and so we must ask whether or not the tests were indeed randomized trials. If they were, that supports the resulting evidence and its implication for the hypothesis under study, and if not, that is a mark against.

Demarcation implies a binary evaluative judgment, while the evaluative procedure recommended here, following insights of Longino (1990) and Solomon (2001), holds that our evaluative attitude to science should be graded. There are many possible justificatory norms for any given domain, and for some unit of scientific work under evaluation, only a subset of them might be satisfied, thus satisfaction of the set of relevant justificatory norms is a graded matter. Moreover, the violation or satisfaction of a particular justificatory norm can itself sometimes be graded. Suppose, for example, that a model has ten empirically underdetermined continuously valued parameters. For each parameter, we can perform a sensitivity analysis (that is, we can vary the value of a parameter and observe the impact on the output of the model), and for each sensitivity analysis, we can explore a wide range of possible parameter values; so, we can satisfy the putative justificatory norm "perform sensitivity analyses on your model" to a greater or lesser degree by performing sensitivity analyses on more or fewer of the parameters, and for each parameter, exploring more or fewer of the possible range of parameter values. For some contexts, the all-or-nothing satisfaction of a particular justificatory norm might be necessary. To use the example from the preceding paragraph, when testing new pharmaceuticals for clinical use, one might categorically require the satisfaction of the justificatory norm "test new pharmaceuticals with randomized control trials." Thus, if a test of a new pharmaceutical were not randomized, our judgment of the test's results would be entirely dismissive or at least seriously downgraded. But, for many contexts, deontic evaluation delivers graded judgment.

Nevertheless, it is perfectly coherent to form categorical evaluative attitudes based on graded properties. We do this with student evaluation, for example. Many institutions assign a numerical score to student performance and then impose a binary judgment on top of that score: A student who scores less than a particular threshold receives a failing grade, and a student who scores above that threshold receives a passing grade. Although the primary evaluative procedure recommended here generally delivers a graded judgment, in some contexts, we might decide to impose a binary judgment, if, for instance, the unit of scientific work under evaluation is extremely compelling or extremely unreliable or if a concrete action must be taken based on the evaluation.

Similarly, one might note that the deontic evaluation recommended here has sacrificed an important aim of earlier demarcation projects, namely, the provision of a basis for coarse-grained pro or con attitudes to wide domains of inquiry. Despite all the problems of Popper's falsificationism, it had at least this much going for it: With one stroke of Popper's pen, nuclear physics of his day was rightly deemed scientific and Freudian psychoanalysis was rightly deemed pseudoscientific. Deontic evaluation is not so neat and simple (though, arguably, Popperian falsificationism was not so simple either—see Grünbaum 1979). Nevertheless, if one aimed to maintain a coarse-grained pro or con attitude to an entire domain of inquiry, one could apply the method suggested in the previous paragraph. One could assess, for a particular domain, the extent to which justificatory norms are satisfied in general over time in that domain, and if, like a failing student, behavior in that domain violates too many relevant justificatory norms, then one can adopt a general negative evaluative attitude to that domain, just as one would assign a failing grade to a mediocre student (one might call that domain "junk science" or "pseudoscience"). It is no mystery that national science funding agencies devote resources to astronomy but not astrology.

Yet dispelling with the coarse-grained evaluation of a wide domain of inquiry is, on the whole, probably for the best. Condemning an entire domain, such as an indigenous healing practice, could shut down dialogue sorely needed in our fractured societies. More fine-grained evaluation enables us to pinpoint precisely where a domain of inquiry lacks sufficient justification and, conversely, allows such domains to articulate reasons, both epistemic and practical, that may be legitimate.

As noted above, the object of evaluation for accounts of demarcation has varied: For Popper it was theories, for Lakatos it was research programs, and for Longino it was the behavior of communities of inquirers, particularly their practices of offering and receiving criticism. Like the object of evaluation in Longino's account, the object of evaluation in deontic demarcation is *action*. Norms state what to do in specific circumstances—they have the form "Do x when in C" (Glüer and Wikforss 2009). The justificatory norms delivered by the Leibniz procedure take this form. The object of evaluation in deontic evaluation is whether a scientist or group of scientists did x when in context C (if satisfaction of x is binary) or the extent to which scientists did x when in C (if satisfaction of x is gradable). In chapter 4, I turn my focus from the evaluation of action to that of scientific assertion.

Recall from §2.1 that the traditional demarcation project was motivated by the question of what beliefs are epistemically warranted. We saw that the single-criterion demarcation principles cannot guide us on epistemic warrant. Deontic evaluation, however, is perfectly suited to this task. The epistemic warrant for belief is very clearly a function of the amount of justification for a belief, and that in turn is a function of the extent to which the justificatory norms relevant to that belief are satisfied.

2.4 Objectivity and Trust

We expect science to be objective. What, though, does it mean for science to be objective? Philosophers have articulated numerous conceptions of objectivity. A common theme in this literature is that there are numerous senses of the notion of objectivity that cannot be reduced to a single notion. Douglas (2004), for example, argues for the "irreducible complexity" of objectivity: Objectivity can be about people's interactions with the world, people's reasoning processes, or social and procedural structures. She parses eight kinds of objectivity within these three broader classes. What Douglas calls "manipulable" objectivity is the capacity of experimenters to manipulate their objects of study and use their objects of study in a range of other contexts. If we can get a scientific result through multiple experimental means, we get "convergent" objectivity. The second broader class of objectivity, focusing on reasoning processes, includes "detached" objectivity, which prohibits wishful thinking; "value-free" objectivity, which prohibits all nonepistemic values from scientific reasoning (see chapter 3); and "value-neutral" objectivity, which occurs when a reasoner is balanced regarding a range of possible value influences. The third broader class of objectivity, focusing on social processes, includes "procedural" objectivity exemplified by quantitative techniques that block the influence of personal idiosyncrasies; "concordant" objectivity, which occurs when there is consensus about a claim; and "interactive" objectivity, which involves the active sharing of results and critical discussion of methods and assumptions akin to Longino's (1990) account of objectivity. Similarly, as noted above, historians of science Daston and Galison (2007) argue that scientific objectivity has meant very different kinds of things throughout history.

In a fun and mildly cantankerous essay, Hacking (2015) pushes back against such scholarly talk about objectivity, advising instead a focus on the details of inquiry that render it more or less trustworthy. Rather than asking questions like "Is this group's research on climate change objective?," we should ask questions like "Is research funded by pharmaceutical companies trustworthy?"—a stance I have some sympathy with. Nevertheless, Koskinen (2020) argues that the diverse conceptions of objectivity discussed by philosophers can be unified, not quite by the notion of trust but by the notion of reliance: When we say that x is objective, we signal that x is reliable, that we endorse x, and that others should too.

Koskinen claims that reliance, in turn, is a function of the extent to which epistemic risks that arise because of our imperfections as epistemic agents are appropriately managed. For example, concealing which subjects are in the intervention group and which are in the control group of a randomized drug trial aims to block the influence of placebo on subjects and minimize any unconscious biases of the researchers. Koskinen's account of objectivity is conceptualized in a narrower manner than is necessary, however, since it is based on epistemic risks that arise only from flaws as epistemic agents and not those epistemic risks that arise from flaws in our methods or instruments.

The deontic approach to the evaluation of science presented so far respects Hacking's advice while also supporting Koskinen's insight that the many meanings of objectivity can be unified, while conceptualizing the notion of objectivity in a more capacious manner than Koskinen does. Scientific objectivity, on the deontic approach, is a function of the satisfaction of justificatory norms. A particular scientific claim is objective, on this account, if and only if the claim is supported by reasons and evidence that result from the satisfaction of relevant justificatory norms. Koskinen rightly argues that objectivity is a gradable property, and so the definition just given can be modified accordingly: A particular scientific claim is objective to the extent that it is supported by reasons and evidence that result from the satisfaction of relevant justificatory norms (one might wish to maintain a threshold for objectivity, whereby a scientific claim is objective as long as it is supported by reasons and evidence that result from the satisfaction of relevant justificatory norms to a sufficient degree).

As we saw in chapter 1, the Leibniz procedure provides the normative basis for the many justificatory norms in science. Some justificatory principles are domain-general and include norms that involve Longino's structured practices of criticism, for example. Some justificatory practices are much more fine-grained and domain-specific, like a norm that tells laboratory technicians to rinse test tubes three times with distilled water. These various justificatory norms are directly pertinent to the various notions of objectivity discussed by historians and philosophers of science. For example, if the laboratory technician does not rinse the test tube three times, the residual matter remaining in the test tube could affect the next experiment, rendering it less objective in Douglas's first broad class of objectivity. Common knowledge as an aim of science requires the provision of justification for public claims and the criticism and defense of that justification, thereby generating the conditions under which Douglas's third broad class of objectivity can be attained. I address Douglas's second broad class of objectivity in chapter 3, where I defend the importance of a value-free ideal for science. To put the point in Koskinen's terms, one of the consequences of satisfying the justificatory norms is the minimization of epistemic risks (in their many guises, not just those that result from our limitations as epistemic agents). Scientific objectivity is a function of the extent to which the justificatory norms recommended by the Leibniz procedure are met. And the Leibniz procedure itself is fundamentally objective in a slightly different sense because it requires the deliberators to put themselves in an intersubjective perspective when assessing putative justificatory norms.

The deontic account of scientific objectivity has an affinity with Longino's (1990) rightly celebrated account of objectivity. The centerpiece of Longino's account, as noted, is based on structured practices of criticism in science. Science has many venues in which criticism is articulated and enforced, such as peer review, hiring committees, and training, and such criticism is an aid to objectivity because it can expose problematic background assumptions in one's empirical methods or conceptual and theoretical apparatus. As discussed in chapter 1, such critical exchanges can be modeled dialogically, and the critical dialogues in Longino's account are first order in that they are exchanges among actual scientists arguing about actual methods, evidence, and theories. The deontic account of objectivity offered here supplements Longino's first-order account with the second-order critical dialogues that occur between the idealized deliberators in the Leibniz procedure, thereby providing a secure normative foundation for a theory of scientific objectivity.

Just as the traditional demarcation project was misguided, so, too, is the current project of asking how to improve public trust in science. As we saw above, one reason the traditional demarcation project was misguided was that it was limited to upholding the same evaluative attitude for all domains or theories properly deemed scientific. While I share the concerns of works such as Oreskes (2019) and Goldenberg (2021), these works invite a response similar to that of the traditional demarcation project—for example, when asking why we should trust science, Oreskes (2019, 56) answers, "When we evaluate the track record of science, we find a substantial record of success." The dialectical situation here is delicate since one can be enormously impressed by the successes of science (as I am) while maintaining that science is too diverse to hold a single overarching evaluative attitude to all its elements. Deontic evaluation permits a finer-grained, localized evaluation. Humanistic scholarship about science should avoid questions like "How can we get the public to trust science?" It is better to ask questions like "What factors contribute to science being more or less trustworthy?" and "How can we improve trust in a scientific domain when that domain is in fact trustworthy?" It is no accident that the motto of the Royal Society is Horace's *nullius in verba*—take nobody's word for it.

A Russian proverb advises, доверяй, но проверяй (when spoken, it rhymes: *doverai, nu proverai*), which translates as "trust, but verify." Oreskes (2019) quotes Ronald Reagan stating the English translation of the proverb for the epigraph of her book *Why Trust Science?* Yet native Russian speakers tell me that what follows the comma is thinly veiled advice *against* what appears before the comma. The proverb might have come from Lenin, perhaps derived from his saying "Не верить на слово, проверять строжайше," which translates as "Do not trust words—check everything strictly." And, indeed, Reagan's statement was in the context of Cold War nuclear arms negotiations, which was clearly no place for trust. The proverb applies just as well to science. Central to science has always been the *withholding* of trust (the first half of Lenin's saying just is the Royal Society's *nullius in verba*) and checking everything strictly, from demonstrations at the Royal Society in past centuries to multilab replication efforts today.

Online venues have recently been created to facilitate careful evaluation of science by other scientists and volunteer sleuths. One is PubPeer, where people can ask questions about and voice criticisms of published articles. The authors of the articles often respond, generating an open discussion of methods and results. Another such venue is Retraction Watch, which tracks retractions of articles from scientific journals and offers commentaries on some prominent retractions. A retraction occurs when a journal formally "unpublishes" an article after it has been peer-reviewed and published. Retractions often result from criticism by other academics or volunteer sleuths who discover problems with an article. A high-profile example occurred with former Stanford University president Marc Tessier-Lavigne, who had previously been chief scientific officer at biotechnology company Genentech. Investigators found that multiple articles he had published in prestigious journals such as *Science* and *Nature* contained manipulated images. Another high-profile example is the retraction of four articles in 2023 because a Harvard Business School professor (Francesca Gino) allegedly manipulated data in studies of . . .

honesty. Retraction Watch includes a list of articles about COVID-19 that have been retracted, and, as of January 1, 2024, the day that I write this (Happy New Year, dear reader), that list includes 388 retracted articles. In total, Retraction Watch now has 43,000 retracted articles in its database. In short, the imperative to check everything thoroughly can be done in scientific venues and extrascientific platforms.

The motivation for what can be called the trust-in-science movement is indeed important. Some widespread beliefs about the world are shockingly unscientific, sometimes dangerously so. Oreskes (2019, 17) notes that in the United States, 67 percent of people who regularly attend church believe that God created humans in the last ten thousand years. Yet consider a different example that pulls in another direction. In *Medical Nihilism*, I noted a study in which one-third of survey respondents agreed that medicine can now cure almost any illness (Stegenga 2018, 8), which represents an absurdly exaggerated trust in medicine. The two examples paint a picture more complicated than what the trust-in-science movement sometimes seems to suggest: The problem is not insufficient trust in science but poorly distributed trust across particular domains and claims of science.

That said, some details of the trust-in-science movement may appear similar to the position offered in this book: Both Oreskes (2019) and Vickers (2023), for example, emphasize the importance of scientific consensus and diversity in science when asking whether science should be trusted. Oreskes, channeling Longino, further stresses the importance of "transformative interrogation" to ground the objectivity of scientific claims. This aligns with the epistemological framework introduced in chapter 1.

The difference between the trust-in-science movement and deontic evaluation offered here is perhaps one of emphasis: Deontic evaluation is primarily a framework for evaluating the reliability of particular scientific claims and, perhaps, particular domains rather than the trustworthiness of the entire edifice of science. Moreover, as suggested in chapter 1, the framework of deontic evaluation is primarily appropriate for scientists themselves since properly assessing the justificatory status of a scientific claim routinely requires the technical, arcane knowledge that comes with years of scientific training. Conversely, for nonscientific laypeople, the position of scholars such as Oreskes and Vickers—that scientific consensus and diversity in science can be signs of trustworthy science—is compelling. Finally, there is a coarse-grained sense in which the prescriptive upshot may also be similar: All people should trust scientific knowledge when such claims to knowledge are contributions to common knowledge, yet that upshot is clearly distinct from a prescription that people should in general trust science. (I have not claimed or argued that scientific knowledge can only be relied on if it is already part of our common knowledge. Far from it. Indeed, we often must and should do with much less, as is widely recognized in the literature on values in science, for instance.)

2.5 Bias

A concept that appears to be roughly the converse of objectivity—and here I argue that in science, it is precisely the converse—is bias. The intuitive connection between these concepts is clear. For example, if a person is biased, it is natural to say that they are not objective. Conversely, if a person is not objective, it is natural to say that they are biased. The notion of bias is, of course, crucial for science. Experiments with biases generate biased and, therefore, misleading evidence, so scientists often do all they can to identify and eliminate or correct for biases.

Kelly (2022) defends a "norm-theoretic" account of bias, which holds that "a bias involves a systematic departure from a genuine norm or standard of correctness." By *genuine norm*, Kelly means not merely statistical norms or social norms but normative norms, norms with evaluative and prescriptive content, like the justificatory norms at the heart of the deontic philosophy of science pursued in this book. The norm-theoretic account of bias faces some challenges. For example, we call a coin that lands heads more than tails "biased," but coins are not governed by "genuine norms," and the fact that most coins land heads and tails at roughly the same frequency is a merely statistical fact (while coin designers and manufacturers might usually aim to produce coins with a symmetric mass distribution, they would not violate any moral or prudential code if they decided otherwise). So not all biases involve systematic departures from genuine norms. Consider another challenge: Suppose that when I am asked to review papers that have been submitted to journals, I always recommend that the journal rejects the paper, thereby violating a plausible norm that suggests evaluating papers on their scholarly merits. I do this systematically, but it is unintuitive to call my behavior biased since I use the same method for any paper, regardless of the topic or who the author might be. So not all systematic departures from genuine norms are biases.

Nevertheless, in science, conceiving of biases as norm violations is promising. As we have seen, a deontic philosophy of science is founded on a set of justificatory norms. Satisfying those justificatory norms renders one's work objective, as we saw in the previous section, while violating those justificatory norms renders one's work biased. Consider the norm that we evaluate new drugs with randomized trials. Violating that norm introduces biases; for instance, if the subjects were not randomly allocated to the experimental groups of the trial but rather were allocated according to some other procedure, say, according to their last name, that would introduce a bias insofar as the groups could end up differing on a possible confounding factor, introducing causally relevant differences between the group receiving the experimental drug and the group receiving the control intervention. So, in short: Biases in science are a result of departures from the justificatory norms delivered by the Leibniz procedure. We can refer to this as the deontic account of bias in science. (Bueter [2022] suggests a related account of bias—what she refers to as an epistemic account of bias—which holds that scientific research is biased if we have good reasons to think that the research could have been performed better.)

The usual way of thinking about epistemic biases is in consequentialist terms; that is, epistemic biases are usually thought to be systematic deviations from reliability or, in other words, systematic threats to truth-conduciveness. Using the example of testing drugs, the bias introduced by a nonrandomized experimental design increases the probability that a confounding factor will skew our estimate of the effectiveness of the drug away from the true value. Yet, according to the deontic account of epistemic normativity defended in chapter 1, not all epistemic biases are merely threats to truth-seeking. Recall that truth-conduciveness was just one desideratum for the deliberators in the Leibniz procedure—though it is, of course, a very important desideratum. Some of the cases that motivated the other desiderata also show that not all epistemic biases should be conceived in terms of a threat to truth-conduciveness. Consider Sasha's method of determining whether a number is a circular prime: The threat to truth-conduciveness is trivial, yet it is intuitive to think that this method is biased (though Sasha's method systematically undercounts circular primes, the degree of undercounting decreases the more she uses her method). More telling is the example of Sara, who systematically manipulated data such that the data indicated the true value of Hubble's constant—Sara's method was plausibly biased, but it was truth-conducive. The deontic account of bias delivers an intuitive verdict about Sara's method.

Kelly notes that we can distinguish a biased process from a biased outcome and argues that neither is generally more fundamental. Therefore, we must maintain both notions to explain the full range of possible biases. Textbook discussions of bias contribute to such a view: Consider a common visual depiction of bias in which arrows have been shot at a circular target and they hit the target some distance from the center in a systematically skewed way (imagine they cluster in the top-right corner of the target). That appears to be a biased outcome. Note, though, that such an outcome could only occur if there was a bias in the process (say, a manufacturing flaw in the arrows or a problem with the archer's vision) or by chance, but that latter explanation is generally improbable, and regardless, chance is conceptually barred as the basis of bias, a point Kelly rightly emphasizes, because chance is neither systematic nor directional (as Confucius says, "When the archer misses the center of the target, he turns around and seeks the cause of failure within himself"). This suggests a general thesis: In the epistemic realm, at least, all there is to a bias is a biased process. Highly suggestive in this regard is that when methodologists articulate lists of epistemic biases, they are all process-oriented (see, for example, the Catalogue of Bias website hosted by the University of Oxford). Another simple, familiar nonepistemic example can motivate this claim. Consider a sequence of coin tosses. On Kelly's distinction, a biased process would refer to anything about the coin (say, asymmetric mass distribution) and its tosses (say, quirky thumb flicks) that cause the coin to land more frequently on one side than another. A biased outcome would refer to a frequency of one side landing more often than the other. But notice that the only possible causes of a biased outcome in this sense are either a biased process or chance. Yet as noted, in Kelly's terms, chance is conceptually barred from being the basis of bias, a point I find plausible. So, the cause of the biased outcome is a biased process. It follows that bias is fundamentally a property of processes and not outcomes. Again, this aligns with the deontic account of bias.

If we were to entertain the distinction between biased process and biased outcome in science, bias in process would be assessed by the extent to which relevant justificatory norms are violated, and bias in outcome would be assessed by the extent to which the resulting particular scientific claims depart from the truth. However, that latter notion of bias is misguided. One problem is that it faces the retrospective benediction argument described in the introduction: In science, the determination of some claim as being true or approximately true can take many years, and so in real time, scientists can only make estimates of truth based on whatever evidence and other epistemic considerations they have access to. Thus in real time, scientists do not have epistemic access to bias in outcome other than their epistemic access to bias in process. More fundamentally, the example of Sara shows that epistemic bias is not a function of the extent to which particular scientific claims are close to the truth because Sara's manipulated evidence rendered her conclusion closer to the truth. Most fundamentally, the epistemic version of the coin-toss argument concludes that a scientific outcome that departs from the truth is due either to the use of biased scientific methods or to chance, but the latter is conceptually blocked as the basis of bias, thus the former is the basis of bias. Thus, bias in science is fundamentally a matter of biased process, itself understood as violations of relevant justificatory norms.

2.6 New Norms: Preregistration

New justificatory norms routinely arise in science as a result of facing new problems, acquiring new technologies, or developing new innovations in methods. Consequently, the evaluation of science should shift accordingly. Earlier I used the example of randomized trials, developed by Fisher in the 1930s (see Hacking 1988 for a fascinating discussion of the history of this development). Fisher developed a new justificatory norm, namely, that if one intends to test an intervention with a population-level experiment, one should randomly allocate the experimental subjects to the experimental groups.

A new norm has recently emerged, particularly in psychology and medicine: Researchers are now strongly urged and often required to preregister their experimental and analytic methods before beginning their research. Registration can occur in journal publications or devoted repositories. This is now one of the most prominently articulated methodological prescriptions in the psychological sciences, yet it was virtually unheard of before the twenty-first century. Placing justificatory norms in their historical and scientific context can be illuminating. Why should scientists preregister their experimental and analytic methods? To put the question in terms discussed in chapter 1: Where does the norm of preregistration get its normative status? Would the Leibniz procedure deliver this norm as a universal, general norm, or would the Leibniz procedure conclude that the status of preregistration as a justificatory norm is localized to specific scientific contexts?

The basic answer to why scientists should preregister their experimental and analytic results is that preregistration guards against what are sometimes (unfortunately) called "questionable research practices," specifically p-hacking (analyzing one's data in a variety of ways to increase the chance of getting a positive or "statistically significant" result) and "harking" (hypothesizing after the results are known, which strikes many as ad hoc). In previous work, I argued that p-hacking and harking are often but not always epistemically pernicious, and I articulated formal conditions under which those practices are, and are not, misleading, and thus the formal conditions under which preregistration renders scientific work more reliable (Hitzig and Stegenga 2020). Yet we can also ask about the substantial conditions under which preregistration is likely to be reliability-enhancing and, conversely, the substantial conditions under which preregistration is not likely to be reliability-enhancing. The history of science can be a guide here since, as I suggested above, a case can be made that there was a scientific golden age from roughly 1890 to 1950 in a wide range of disciplines, yet preregistration was not yet a norm, which is highly suggestive that, in principle, good science in general does not hold preregistration as a norm, though in particular circumstances it may.

Nguyen's (2023) insightful notion of "hostile epistemology" is helpful to make sense of this. Hostile epistemology emphasizes the influence of a context or environment on the capacity of agents to make reliable inferences about the world. To address the epistemic status of preregistration, we can distinguish two extreme contexts: an epistemically *ideal* environment and an epistemically *hostile* environment. An epistemically ideal environment has the following three features.

Assertions are modest. In epistemically ideal environments, speakers make claims modestly, and those claims are adequately justified and appropriately hedged; two examples are the title of James Chadwick's 1932 article "Possible Existence of a Neutron" and the title of a 1934 article by Enrico Fermi: "Possible Production of Elements of Atomic Number Higher than 92." Both claims were sufficiently humble, despite the momentous nature of the discoveries (both Chadwick and Fermi were awarded Nobel Prizes for the research that went into those articles).

Methods are transparent. In epistemically ideal environments, scientists have epistemic access to each other's methods. When making a claim, scientists fully articulate the justification for that claim, which includes a faithful description of one's methods. As noted earlier, I am impressed by Merton's ability to write in 1942 that he had observed a near-total absence of fraud in science, suggesting a high degree of transparency in some areas of science at the time.

Abundant low-hanging fruit. In epistemically ideal environments, many important truths are available to be discovered through methods that, while often very clever, are imminently feasible. Again, atomic physics in the early decades of the twentieth century exemplifies this feature. Consider the Geiger-Marsden-Rutherford gold foil experiments, which involved very basic tools and led to the discovery of the dense nuclear core of atoms, which was one of the first of many epistemic fruits rapidly picked in those decades in atomic science.

Conversely, an epistemically hostile environment has the opposite features.

Assertions are bold. In epistemically hostile environments, speakers make bold claims and often with little justification. In chapter 4 I describe an example in which the scientists responsible for the Bangladesh mask study wrote an editorial in a major newspaper titled "We Did the Research: Masks Work," even though the positive evidence for the claim was extremely thin.

Methods are opaque. In epistemically hostile environments, scientists do not have epistemic access to each other's methods and do not give full accounts of their methods when publicly justifying their claims. This is one of the key problems with p-hacking since the unconstrained analytic choices made during this process are typically not publicly reported.

Rare fruits, hard to reach. In epistemically hostile environments, significant truths are harder to come by, perhaps because there are fewer such truths in the domain or perhaps because they are more difficult to discover. Bird (2021) argues that the replication crisis in psychology is a result of there being few hypotheses in psychology with a relatively high prior probability, which is a version of the rare fruits aspect of an epistemically hostile environment.

If a scientist engages in p-hacking or harking in an epistemically ideal environment, an agent who reasons well can observe the scientist's methods (because they are transparent) and notice that the scientist uses methods that are likely to increase the probability of generating spurious confirmation for a hypothesis, and so she can adjust her credence in that hypothesis accordingly (Hitzig and Stegenga 2020). Moreover, since the scientist asserts modestly, the agent is not misled. Thus, in an epistemically ideal environment, practices like p-hacking and harking need not be misleading. Finally, because many important truths are feasibly discoverable, there is a nonnegligible probability that the hypothesis in question is one of those truths, and thus subsequent research is likely to replicate this work. So, in an epistemically ideal environment, preregistration confers little benefit. Conversely, in that same environment, preregistration may be costly. Scientists who are constrained in their work by a preregistration plan are more likely to miss findings they might otherwise discover if they were free to explore.

However, if a scientist engages in p-hacking or harking in an epistemically hostile environment, agents cannot learn specific details of a scientist's methods because those methods are opaque. Indeed, a scientist may actively mislead others about the details of the methods used. So an agent cannot coordinate her credence to reflect the fact that the scientist engaged in p-hacking or harking. Moreover, since the scientist asserts boldly in this environment, agents are liable to be misled. Finally, because there are few important truths that are readily discoverable, there is a nonnegligible probability that the hypothesis being entertained is false, and thus further research is unlikely to replicate this work. So, in an epistemically hostile environment, preregistration can confer benefits: At the very least, preregistration can function as a transparency-enforcement mechanism, thereby giving agents more visibility about the methods used and thus a chance to properly coordinate their credences.

This analysis explains why preregistration has become so important in domains such as psychology and clinical research, because such domains are plausibly closer to being epistemically hostile environments. Conversely, it also explains why preregistration was not needed in domains like atomic physics in the early twentieth century, because that domain was plausibly closer to being an epistemically ideal environment. We can understand the emergence of this new norm by looking closely at the contextual details of epistemic environments. More generally, this example illustrates the importance of a granular approach to scientific evaluation, as is possible with deontic evaluation.

2.7 Summary

In §2.1, I noted that the leading approaches to scientific demarcation in the twentieth century were based on a single property, were product-oriented, and delivered an ungraded assessment. The most famous of such approaches to demarcation was Popper's falsificationism. Problems with existing approaches to demarcation have led many philosophers to abandon the project altogether, though the initial motivation of the project—to provide guidance on belief and action in light of scientific claims—is clearly as important as ever. In this chapter, I offered a new approach to scientific demarcation by first noting that the demarcation project should be considered a form of evaluation of science; thus the aim of the demarcation project can be met by a compelling account of the evaluation of science. The resources developed in chapter 1 provide such an account. The evaluation of any unit of science—empirical work or theoretical product—should proceed by assessing the extent to which the relevant justificatory norms have been met. To assess the health of science, we must assess the heart of science. A deontic philosophy of science affords new accounts of bias, objectivity, and conditions for trustworthy science.

CHAPTER THREE

A New Value-Free Ideal

3.1 Introduction

Ancient Indian philosophical texts from both the Vedic and Buddhist traditions describe the ultimate good for a person as the complete dissolution of all egotistical attachment and the attainment of a mental state free of all desire, hope, anxiety, pleasure, and pain. If this state is meant to represent an ideal, then it is presumably a desirable state to attain. However, an austere state of complete detachment does not seem very attractive. If a person has eliminated their desires for and experiences of pleasure, it is hard to see how that state could be considered subjectively good for that person. How can a state be good for us if it renders us impervious to either positive or negative influences?

The Indian philosophical tradition has developed a variety of responses to this puzzle. Ganeri and Adamson (2020, 8) suggest that the appeal of this state of complete detachment comes not from the benefit of attaining it but from the benefit of pursuing it:

> Perhaps the very fact that the ideal states are described in such unappealing terms shows us that these are not really intended as descriptions of the good for human beings. We should instead ask: how does the idea of striving to achieve such a state help one make progress? ... If I genuinely believe that the ideal state involves no pleasure at all, I am apt to allow myself to be nourished by the pleasures I do have without being distracted from my other goals by the need to seek out new pleasures. In other words, I may come to lead a life of restraint and self-control.

This is interesting because it suggests that attaining the idealized end-state is not itself good, but actively pursuing that state is good. In terms

I explain as this chapter proceeds, this response decouples pursuit desirability from end-state desirability. Reiterating the emphasis on process over product in deontic philosophy of science, I deploy this decoupling of pursuit desirability from end-state desirability to develop a novel value-free ideal for science.

The value-free ideal for science holds that scientific reasoning should not be influenced by nonepistemic values, that is, social, cultural, political, or ethical values that are not relevant to the truth of scientific hypotheses. The value-free ideal is, at first glance, extremely plausible. One might even think it hardly needs philosophical defense. Science is our best source of objective knowledge about the world, and the influence of nonepistemic values on science threatens such objectivity. Despite its appeal, however, the value-free ideal has faced several philosophical challenges, leading to a new dogma in philosophy of science that rejects the value-free ideal as both unattainable and undesirable. This chapter aims to articulate a new version of the value-free ideal that avoids the existing philosophical challenges.

The core argument is that the feasibility and desirability of attaining an end can be decoupled from the feasibility and desirability of pursuing that end, as illustrated in the example above from the Indian philosophical tradition. A particular end-state may be unfeasible or undesirable, yet pursuing that end may nevertheless be feasible or desirable. The ideal of world peace may be impossible to achieve, yet the pursuit of that ideal is both possible and good; de-escalating conflicts, disarmament, and a more equitable distribution of resources are all potential ways to approach world peace, and these are, at least to some degree, feasible to enact. The decoupling strategy is trickier for desirability. It may seem odd to suggest that it is good to pursue an end that one judges undesirable, but as we have seen with the initial example, this is a serious possibility for at least some contexts. It is obviously true that some means to a particular end can be desirable even if the end itself is not. A state of complete starvation is undesirable, but cutting junk food from one's diet may be good, even though it is a means to the end of starvation. What is at stake in the upcoming argument is stronger: not merely the desirability of taking certain actions that happen to be a means to an undesirable end but the desirability of actively pursuing an undesirable end itself. Challenges to the value-free ideal have focused on the feasibility and desirability of a value-free end-state. Yet one can grant that the value-free ideal is neither end-state-feasible nor end-state-desirable while arguing that the *pursuit* of value-freedom is both feasible and desirable. That is my goal.

I start by articulating the primary arguments in favor of the value-free ideal and the main challenges to the ideal (§3.2). I then argue in §3.3 that the pursuit of the ideal is feasible by striving to attain common knowledge and specifically by satisfying the justificatory norms delivered by the Leibniz procedure described in chapter 1. In §3.4, I argue that the pursuit of the value-free ideal is desirable even if the end-state is undesirable. The conclusion is a specific and novel version of the value-free ideal, which holds that scientists should act as if science should be value-free. This value-free ideal exemplifies the deontic approach to philosophy of science twice over since this value-free ideal is itself a deontic principle and, as we will see, it can be pursued by striving to satisfy the deontic justificatory norms delivered by the Leibniz procedure.

3.2 The Value-Free Ideal

The value-free ideal is a somewhat nebulous notion and has been interpreted in several ways (Proctor 1991; Lacey 1999). It is uncontroversial that values influence the choice of scientific research projects, constraints on research methods, and applications of scientific results. The issue under dispute in recent philosophical literature is whether values influence the internal structure of scientific reasoning or the inference from evidence to conclusion. The value-free ideal only prohibits values that play a role in this internal, inferential stage of science.

Another standard caveat: Science involves ampliative inference, so if one wants to distinguish between better and worse inferences, one must appeal to at least some normative considerations that go beyond mere deductive validity. The values that contribute to the achievement of the epistemic aim of science are standardly referred to as *epistemic values*. These values specify features of scientific theories—examples include simplicity, empirical adequacy, and fruitfulness—to be used as a basis for judging how a theory fares relative to its rivals on epistemic grounds. In terms of chapter 1, a value is epistemic only if it contributes to scientific justification. Indeed, epistemic values can be the basis of justificatory norms. Consider the simplicity of theories. This has been a topic of debate, but if we suppose for the sake of illustration that theoretical simplicity is indeed an epistemic value, a corresponding justificatory norm could be, roughly, "All else being equal, formulate theories that are as simple as possible." In short, the value-free ideal does not prohibit epistemic values;

rather, it prohibits those values that do not serve an epistemic function, such as social, moral, and political values.

The caveat in the preceding paragraph relies on a distinction between epistemic and nonepistemic values. The Leibniz procedure described in chapter 1 provides a basis for this distinction. For any putative epistemic value, we must ask if the ideal deliberators in the Leibniz procedure would formulate a justificatory norm based on the value; to determine that, as we saw, the ideal deliberators appeal to the desiderata of truth-conduciveness and manifestation and promotion of epistemic agency and responsibility.

Taking the above qualifications into account, here is a preliminary characterization: The value-free ideal holds that nonepistemic values ought not influence the internal features of scientific reasoning. Yet the notion of "influence" is ambiguous, as is the notion of "internal features of scientific reasoning." To further clarify, I introduce the notion of a *bifurcation point*: If evidence E in conjunction with values V_1 justify conclusion C_1, and E in conjunction with values V_2 justify a conclusion C_2, and C_1 and C_2 are incompatible, then there is a bifurcation point between V_1 and V_2. Bifurcation points constitute value-influence on scientific reasoning. This is the "difference-to-inference" model for values in science (Stegenga and Menon 2023). The value-free ideal defended here holds that scientists should strive to eliminate bifurcation points. This is intuitive since the existence of a bifurcation point entails that the justificatory status of a scientific finding is influenced by nonepistemic values, yet nonepistemic values are by definition irrelevant to the justificatory status of a hypothesis. Moreover, the existence of a bifurcation point regarding a hypothesis entails that the hypothesis cannot be common knowledge, because a bifurcation point implies disagreement, which entails an absence of strong consensus about that hypothesis, and thus regarding that hypothesis the constitutive aim of science has not yet been met. Given a fixed evidential basis, whether a scientist regards a hypothesis as true or false should not depend on the values they endorse. The value-free ideal developed here holds that scientists should aim at the elimination of all bifurcation points, and progress toward the ideal can be made by reducing the number of bifurcation points.

If any conceivable value set can contribute to a bifurcation point, then there is plausibly an infinite number of bifurcation points for every scientific inference. Thus, if progress toward the value-free ideal is understood in terms of reduction in the number of bifurcation points, then there must be some restriction on the kinds of values that are relevant to defining bifurcation points. Here I restrict bifurcation points to those points of contention that result from actual value disagreements.

A bifurcation point must involve incompatible conclusions drawn from the same evidence, due to differing values. It is not enough that the conclusions are merely different from one another, even if that difference is attributable to a difference in values. Mere differences in conclusions based on different values are insufficient to conclude that values have influenced the internal features of scientific reasoning because value-based differences in research focus may lead to different conclusions simply because the researchers care about different aspects of the evidence or different fine-grained features of the phenomenon under investigation. The value influence in such cases would be due to external features of scientific reasoning—the research questions being asked—rather than internal features of scientific reasoning. So, in short, my conception of value-freedom allows nonepistemic values to play some role in determining how a conclusion is framed—what terminology is used or what aspects of the evidence are highlighted, for instance—but compels scientists to attempt to eliminate the impact of values on the putative justification of a conclusion.

The difference-to-inference model provides a precise account of the notion of values influencing internal features of scientific reasoning, and, as we will see, provides the basis for a novel, value-free ideal.

Arguments for the Value-Free Ideal

There are several considerations that support the value-free ideal. First, value-permeation threatens the reliability of science (some scholars take this to be controversial; I challenge their arguments below). As argued in chapter 1, the constitutive aim of science is common knowledge, which requires the discovery of truths. Some epic failures in the history of science can be seen as caused by violations of the value-free ideal. Genetics and agricultural biology in the Soviet Union were disastrously harmed by biologist Trofim Lysenko's rejection of Mendelian genetics and support of Lamarckism, based on his belief that the latter was more consistent with Soviet principles. Until the 1970s, primatology was dominated by men, which resulted in an implicit androcentric value framework that biased the interpretation of primate behavior, leading to erroneous theories about female reproductive strategies and social hierarchies. Resistance to advances in the historical sciences during the eighteenth and nineteenth centuries—such as the discoveries in geology suggesting that the earth is much older than biblical sources claim and Charles Darwin's formulation of evolutionary theory—were based on religious commitments. I argued recently that values biased the study of putative sex differences in sexual desire (Stegenga 2022b). (For further discussion of some of these cases, see Elliott [2017]). Such cases demonstrate that science's constitutive aim is hampered when values influence scientific reasoning: Value-permeation necessarily blocks the achievement of common knowledge since the existence of a bifurcation point means that there is no consensus about the hypothesis in question.

Second, value-free science can more justly inform public policy. Since science is often harnessed for policy, and democratic ideals entail that policy should reflect the values of the broad citizenry rather than those of a narrow group of scientific experts, science (especially policy-oriented science) should be value-free (Betz 2013; Bright 2018), or, if value influence is inevitable, science should be influenced by representative democratic values rather than the idiosyncratic values of particular scientists (Schroeder 2021). If values influence the internal features of scientific reasoning in policy-relevant research, a narrow and nonrepresentative set of values can subtly and inappropriately influence policy.

A third and related argument for the value-free ideal is that public support and public trust in science are influenced by the extent to which the public views science as value-free (Bright 2018). The value-free ideal should be upheld as a regulative constraint so that the public supports and trusts science.

Challenges to the Value-Free Ideal

The putative merits of the value-free ideal have been contested, and both its feasibility and desirability have been challenged.

One challenge is that some concepts used in framing scientific hypotheses are irreducibly value-laden. Choosing to use a particular concept when describing, analyzing, or explaining data may indicate certain value commitments, and others who do not share those values might therefore reject a conclusion framed in those terms. Alexandrova (2018), for example, discusses the indispensability of "thick concepts" in some domains of science, which has long been noted in philosophy of science. Anderson (2004) argues that values play a role in framing questions in research about divorce. Atkinson (1998) argues that when measuring poverty rates, value-influenced choices must be made when defining and operationalizing poverty—for instance, the choice between relative and absolute standards of poverty may be influenced by whether one considers it worse to be below a certain percentile of household income or to be unable to afford a particular bundle of goods. If making such value judgments is genuinely unavoidable in some areas of research, then the value-free ideal is unattainable. This is the *framing problem* for the value-free ideal.

Differences in framing will only count as bifurcation points (and, hence, on my characterization, genuine examples of values influencing the internal features of scientific reasoning) if they lead to contradictory conclusions about the same hypothesis. Suppose Aisha, motivated by her set of values, chooses to measure absolute poverty, while Bheem chooses to measure relative poverty. Suppose further that Aisha estimates the poverty rate in a region to be 12 percent, and Bheem estimates it to be 7 percent. If they were evaluating a hypothesis like "the poverty in this region is greater than 10 percent," they would be in apparent disagreement, but there would not be a bifurcation point because they relied on two distinct concepts of poverty. There would be no bifurcation point unless both Aisha and Bheem believed they were addressing the same hypothesis using the same concept, namely, poverty. Only if they drew contradictory conclusions that are formulated with this single concept would there be a bifurcation point. Since absolute and relative poverty are distinct concepts, the apparent disagreement between Aisha and Bheem would not be contradictory, thus there would be no bifurcation point, and hence no value-permeation (for a more detailed version of this objection to the framing problem, see Stegenga and Menon 2023).

A second challenge is what Betz (2013) refers to as the *methodological critique*. One version of this challenge holds that evidence underdetermines support for hypotheses, and values must fill the gap between evidence and hypothesis (Longino 1990, 2004; Elliott 2011). Another version posits that decisions to accept or reject hypotheses are always uncertain, and since errors have practical consequences, our valuations of those consequences influence our decisions to accept or reject hypotheses (Rudner 1953; Douglas 2000). This is known as the argument from inductive risk. The idea is that accepting or rejecting a hypothesis based on a given set of evidence depends on deciding whether the extent to which the evidence supports the hypothesis surpasses a particular threshold. There are two kinds of errors one could make: rejecting a true hypothesis because one's evidential threshold is too high or accepting a false hypothesis because one's threshold is too low. The evidence itself cannot tell us where to set the threshold—that must be determined by an evaluation of the relative consequences of these two types of errors, and values must and should play a role in this evaluation. Therefore, a completely value-free science is infeasible. The infeasibility of the value-free end-state has been argued for in numerous contexts, including vaccine safety (Goldenberg 2021), enzyme classification (Conix 2020), the science of well-being (Alexandrova 2018), epidemiological modeling during the COVID-19 pandemic (Winsberg et al. 2020), and especially climate science (Havstad and Brown 2017a; Frisch 2020; Winsberg 2012). (For a formalization of the argument, see Brown and Stegenga [2023]).

Steel (2016b) points out that the argument from inductive risk can be interpreted as making the descriptive point that values *do* play a role in determining the threshold of hypothesis acceptance or as making the normative point that values *should* play this role. The descriptive reading of the inductive risk argument challenges the feasibility of the value-free ideal, while the normative reading challenges the desirability of the ideal.

Douglas (2000) maintains that the argument from inductive risk applies not just to the conclusions drawn from evidence but also to the characterization of the evidence itself. My definition of value influence might seem unrelated to the characterization of evidence because that definition is based on bifurcation points in the inference from evidence to conclusions. This may appear insufficient to represent a value-permeated dispute about the evidence itself. However, what counts as the evidence may be contextually determined: What is a shared body of evidence in one context may be a disputed conclusion in another. In cases where there is disagreement about the characterization of the evidence, there could be a more basic or straightforward description of the evidence that all parties agree on. For instance, Douglas (2000) describes how different teams of pathologists disagreed about the extent of toxicity visible in rat liver slides. But there could be another description of those slides that did not mention toxicity, about which the teams would agree. Their disagreement could then be framed in terms of what conclusions about toxicity are appropriately inferred from these shared descriptions. So, although this example might be seen as one where there is disagreement about the characterization of evidence, it could also be described as one in which there are bifurcation points in the inference from evidence to conclusions about toxicity.

Another challenge to the desirability of the ideal, prominent in feminist philosophy of science, says that claims of value-freedom usually disguise the unexamined dominance of a narrow range of hegemonic values.

This is the *value-enrichment critique* of the value-free ideal. According to this challenge, we should enrich the range of values that influence science to avoid the biasing effect of a narrow set of dominating values. Longino (2004), for example, argues that complex subjects of scientific study (such as human behavior) are unlikely to be fully understood using a single approach, so, in such domains, the involvement of diverse value perspectives enhances our scientific understanding and mitigates the influence of extant biasing values. (For further discussion of this critique, see Intemann [2005]; Kourany [2008]; and Hicks [2018].) Since scientific reasoning is inevitably value-laden, those values should be sufficiently diverse and representative of a broad range of interests. Therefore, rather than purifying scientific reasoning from values, we should be enriching scientific reasoning with even more values. From this, it follows that the value-free ideal is not desirable. Longino (2004) states the value-enrichment critique explicitly, and Anderson (2004) offers a detailed case that is framed as highlighting the merits of value-enriched scientific reasoning. In §3.5, I offer a reframing of the value-enrichment critique that undermines it as a challenge to the value-free ideal.

Reformulating the Value-Free Ideal

My aim in this chapter is not to deny the significance of the criticisms of the value-free ideal described above. Those criticisms highlight important considerations that any sober view of science must consider. My defense of the value-free ideal does not rely on denying the conclusion they arrive at, namely, that an end-state of value-freedom is neither feasible nor desirable. My strategy is instead to argue for a reformulation of the ideal that makes it clear why pursuing value-freedom is both feasible and desirable, even if the end-state is neither attainable nor desirable. The criticisms of the ideal should not lead us to abandon value-freedom completely but should instead encourage us to rethink what it means for value-freedom to be an ideal for science. In short, existing challenges to the value-free ideal target a stative ideal—which maintains that the state of science at any particular time should be value-free—while my value-free ideal is active, which maintains that scientists should constantly strive over time to eliminate value-influence. This dynamic aspect of my value-free ideal thus avoids the existing challenges.

I will elaborate on the argument in subsequent sections. Here I lay some preliminary groundwork by distinguishing four conceptions of the value-free ideal. Three of these four conceptions are either straightforwardly refuted or rendered implausible by the arguments of the critics of the ideal, but the fourth conception—which is, as far as I can tell, a novel version of the value-free ideal—can be maintained even if one fully accepts the critics' conclusions.

First, let us dispense with the first three formulations. The first is:

VFI1: SCIENCE IS, AS A MATTER OF FACT, VALUE-FREE. This version might seem so obviously wrong that it is not worth discussing. After all, philosophers of science have presented many case studies where values have played a role in scientific inference. It would be naive to maintain that values play no role in science, given the many examples of value-permeation described by critics of the value-free ideal. However, as Lacey (1999, 1) mentions, value-freedom could perhaps be understood as an idealization of science. No doubt the messy reality of research often involves value-permeation, but our best science might permit a rational reconstruction purged of value-permeation. Even if values are involved in the context of discovery, in the context of justification, we may be able to arrive at the same conclusions without an appeal to values. So, while VFI1 may not be strictly true, one might think that it could be true as an idealized reconstruction of science.

However, this interpretation of VFI1 is not defensible. Worries about the ideal cannot be dispelled by appealing to the context of justification because these are worries about the context of justification itself. The critics discussed above note an inferential gap between evidence and conclusion that cannot be filled solely with epistemic values, so nonepistemic considerations are required. A rational reconstruction that defuses this criticism would have to show that one can reason from the same evidence to the same conclusion without an appeal to nonepistemic considerations. This would be possible only if either the inferential gap did not actually exist or if there were additional epistemic values that were not considered in the context of discovery. While there may be some cases that could be understood in one of these two ways, it is not plausible to think that this is a general diagnosis of all or even most cases where there is an inferential gap.

One may recognize that science, even in an ideal reconstructed form, is not value-free but still argue that scientists should not concern themselves with any value-based considerations while engaged in scientific reasoning. Researchers, when considering scientific inferences, should restrict themselves to evaluating epistemic reasons and ignore the intrusion of nonepistemic reasons. This is another potential formulation of the value-free ideal:

VFI2: SCIENTISTS SHOULD ACT AS IF SCIENCE IS VALUE-FREE. There is little to support this conception of the ideal. If there are values involved in scientific reasoning, then ignoring them does not serve the benefits of value-freedom discussed above. Ignoring nonepistemic considerations would not make science more truth-apt, nor would it help fulfill the democratic ideal that policy-oriented science reflects values of the citizenry. It is possible that VFI2 might serve the third benefit of value-freedom discussed above—public trust in science—because scientists claiming value-freedom might convince the public even if the claim is false. However, public trust in science predicated on such an egregious misrepresentation of scientific practice is not worth having. As argued in chapter 2, trust in science should be based on a science that is genuinely worth trusting, not based on a false image of science.

VFI1 and VFI2 are the versions of the ideal that seem to be the primary target of criticisms raised by feminist philosophers of science. They point out that if scientists ignore the role of values in scientific inference, the consequence will be an entrenchment of the values of a socially dominant class since dominant values usually serve as the invisible default in discourse that is not critically examined. To combat the bias inherent in prioritizing a single set of values, philosophers like Longino recommend an explicit recognition of the influence of values. That is compelling—science is not, as a matter of fact, value-free; the purported value-freedom of science can hide the hegemony of an invisible value scheme that is taken for granted, and the appropriate response should be the acknowledgment and mitigation of the dominance of that value framework.

I have rejected the conceptions of the value-free ideal that present scientific practice as value-free, either as a matter of fact or as a pretense. A more plausible alternative is to think of the ideal as articulating a goal toward which science should be directed, rather than a description of current science:

VFI3: SCIENCE SHOULD BE VALUE-FREE. VFI3 does not deny the value-permeation of scientific practice, nor does it claim that scientists should ignore the role of values. It expresses the desirability of a value-free end-state for science. One might assume that arguments against the feasibility of this end-state would derail VFI3. After all, it is a plausible principle that "ought implies can." If it is impossible to attain value-freedom, how could science be given the imperative to be value-free? But that argument is hasty. The "ought implies can" principle, if it is sound, applies to action—the principle holds that if it is impossible for an agent to act in a particular way, then the agent cannot be obliged to act that way. The principle does not preclude the evaluation of unattainable states. Even if complete world peace is unattainable, one can still argue for the desirability of the state. Along similar lines, one may project value-freedom as a regulative ideal that can guide the action of scientists even if the ideal cannot be attained.

This version of the value-free ideal is stative, as it is based on a characterization of a particular state of science as being ideal. However, as noted above, critics dispute the claim that the end-state of value-freedom is desirable, which would be a direct repudiation of VFI3. These critics present arguments for the undesirability of the ideal that are targeted at the ideal itself, because a fully value-free science would have characteristics that would undermine the practical significance of scientific inquiry. I will discuss these arguments at greater length below. For now, let us assume that the value-free end-state may well be undesirable, and if it is, then VFI3 is false.

Nevertheless, a fourth formulation of the value-free ideal is available:

VFI4: SCIENTISTS SHOULD ACT AS IF SCIENCE SHOULD BE VALUE-FREE. Unlike VFI2, this formulation does not require pretense about the value-freedom of current science. VFI4 is compatible with acknowledging the value-permeation of science. Unlike VFI3, this version of the value-free ideal is not committed to the desirability of the value-free end-state. It merely says that scientists should act as if the end-state is desirable. It can thus sidestep the arguments that critics raise against the desirability of value-freedom. It is based on the distinction between end-state desirability and pursuit desirability and the idea that these two notions can come apart. Even if an end-state is not desirable, it is still possibly desirable to pursue that end-state, as we saw with the example in §3.1 from the ancient Indian tradition. Unlike VFI3, VFI4 is not stative, but rather, is active: This version of the value-free ideal is constituted by ongoing guidance for the actions of scientists over time.

Some may argue that even though VFI4 does not require the same sort of pretense that VFI2 demands, it still seems to call for something resembling pretense. This conception of the value-free ideal holds that scientists must act as if a value-free end-state is desirable even if it is not. While this is true, the problem with VFI2 is that the pretense is pointless and likely harmful—acting in the manner recommended by VFI2 will not help fulfill the purported ends of value-freedom. This is not the case with VFI4, as I argue in §3.4. I argue that this conception of the value-free ideal not only evades the criticisms against the value-free ideal, it also guides scientific practice in a way that promotes the benefits of value-freedom. First, though, I argue that this version of a value-free ideal is feasible to pursue.

3.3 Pursuit Feasibility of the Value-Free Ideal

The methodological critique and the framing problem challenge the feasibility of a value-free end-state. However, ideals can fail to be end-state-feasible yet be pursuit-feasible. Even if the state described by the ideal is unattainable, there might be strategies available to approach the ideal. In this section, I argue that scientists can pursue value-freedom, and in the next section, I argue that it is good for scientists to pursue value-freedom.

Values may play a direct biasing role in scientific reasoning, wrongly influencing the degree of apparent justification of a scientific hypothesis. For instance, the influence of values may lead researchers to consciously or unconsciously cherry-pick the data they regard as salient or to collect data in a manner that favors a particular conclusion. But there are ways to eliminate or reduce this biasing effect, specifically by adhering to the deliverances of the Leibniz procedure (described in chapter 1). Methodological strategies such as the random assignment of subjects in an experiment and concealing the allocation of subjects to experimental groups have removed some bias-prone choices from the researchers (*bias reduction*), and the introduction of diverse value perspectives in a research discipline can expose and mitigate implicit and unacknowledged biases (*bias neutralization*).

Values can also indirectly affect scientific inference, by influencing the level of certainty needed to accept or reject a particular hypothesis. Here, too, there are strategies for moving toward value-freedom. If the degree of justification of a hypothesis lies between the evidential thresholds recommended by two conflicting value perspectives (and thus a bifurcation point exists), then researchers could gather more evidence until the justification is clearly above or below both thresholds, allowing supporters of both perspectives to agree on whether to accept the hypothesis (*evidence strengthening*). This would eliminate at least one bifurcation point. Alternatively, one could defer the dichotomized acceptance or rejection of the hypothesis to a later decision-making stage outside the scope of scientific inference (*deferral*). This could be achieved by directly reporting degrees of justification or degrees of certainty for various hypotheses, either precisely quantified or vaguely qualified, rather than committing to one specific hypothesis as the conclusion of scientific inference. Or one could use the strategy discussed in §1.3, in which Aisha and Bheem disagree about C_1 and C_2 but agree on a more coarse-grained claim C_3, which is entailed by both C_1 and C_2 (*hedging*). We will see below, however, that these strategies for eliminating bifurcation points come with costs.

Strategies that mitigate (but do not necessarily eliminate) the influence of values on scientific reasoning—by reducing empirically or theoretically underdetermined methodological choices, by strengthening evidence, by deferring decisions to accept or reject hypotheses, or by hedging—show that bifurcation points can be eliminated, and thus steps can be taken toward a state in which science is value-free. Such strategies are routine aspects of science. Therefore, the value-free ideal is pursuit-feasible.

In the remainder of this section, I discuss the strategies of evidence strengthening and deferral in more detail, and then I defend my novel version of the value-free ideal in §3.4.

Evidence Strengthening

In some circumstances, scientific conclusions can be regarded as beyond reasonable doubt when the evidence for such conclusions is extremely strong. In such a scenario, different value perspectives entailing different thresholds for accepting the conclusion would not necessarily result in bifurcation points because the evidence could be strong enough to surpass all actual thresholds of acceptance. This is a case in which different values do not lead to different conclusions—let alone incompatible conclusions—making the inference value-free in the sense described earlier.

The evidence-strengthening strategy can be illustrated by comparing observational and experimental trials. Evidence from an observational study assessing a correlation between A and B might not differentiate between the hypotheses that A causes B, that B causes A, or that a common cause C causes both A and B. Scientists with different values, and correspondingly different evidential thresholds for the various competing hypotheses, may differ in the conclusions they draw from the study. The evidence-strengthening strategy would recommend performing an experimental trial: If A and B are correlated after A is administered to a randomly assigned group and a comparison intervention is administered to a control group, then evidence supporting the hypothesis that A causes B is considerably strengthened. This makes it harder for values to influence the choice of hypotheses. Even if my values warn strongly against the risk of wrongly concluding that A causes B, leading me to set a high evidential bar for that conclusion, the results of a well-conducted experiment might cross that bar. It is, of course, not impossible for values to influence hypothesis choice even after evidence strengthening—no trial is perfect and the possibility of epistemic error remains, and thus the methodological critique continues to apply—but the conclusion aimed at here is not that value-freedom is attainable but rather that its pursuit is feasible. Steps can be taken toward eliminating the influence of values on scientific reasoning.

To put this in terms of chapter 1: The evidence-strengthening strategy involves greater deployment of justificatory norms. Whatever the Leibniz procedure would recommend as justificatory principles and practice for some context, the evidence-strengthening strategy to eliminate bifurcation points involves further application of those justificatory norms.

Deferral and Hedging

Betz (2013) notes that the methodological critique relies on the premise that generating policy-relevant scientific results requires making decisions regarding the acceptance or rejection of hypotheses that are not fully determined by empirical constraints. Betz rearticulates a response from Jeffrey (1956), which denies that scientists must accept or reject hypotheses and argues instead that they should characterize their uncertainties and report their findings accordingly, leaving questions of acceptance or rejection of findings—or, rather, acting on the basis of such acceptance or rejection—to policy-makers. Rather than reporting their conclusions in terms of "plain hypotheses" (that such-and-such is the case), scientists ought to report their conclusions in terms of "hedged hypotheses" (given this evidence, the probability of such-and-such is so-and-so). This would shift assessments of the sufficiency of evidential support for acceptance or rejection of hypotheses outside of scientific reasoning and into policy deliberations.

The deferral strategy can also help to mitigate the framing problem. As discussed above, Atkinson (1998) argues that the study of poverty involves choosing between value-laden conceptualizations. In such cases, hedged conclusions might help mitigate the influence of values. An economist might be motivated by value-based considerations to convince others that poverty rates during the 1980s were greater in the United Kingdom than in France, and that would be supported by some empirical data with a particular operationalization and definition of poverty. However, with the development of more nuanced ways of conceptualizing and operationalizing the measurement of poverty, one could observe that the economist is relying on particular choices, while on other choices, it could appear that poverty in the United Kingdom was less than that of France. The articulation of alternative conceptualizations shows that the reliance on a single conceptualization and measurement may not be robust. In light of this, scientists should hedge their inferences and claims accordingly.

A challenge to the hedging strategy as a response to the methodological critique is that the evidential sufficiency problem arises again when quantifying uncertainty (Rudner 1953; John 2015; Frisch 2020). Different value sets may disagree about the threshold of evidence necessary to assign a particular degree of certainty to a given hypothesis. In response, Betz uses the hedging strategy to determine degrees of certainty. The idea is to choose a representation of uncertainty imprecise enough such that all contending value perspectives agree that it crosses the threshold of sufficient evidence. If there is contention about point probabilities assigned to hypotheses, it might be better to use intervals instead. If intervals are controversial, qualitative ascriptions of uncertainty might not be; if even that fails, perhaps a simple enumeration of serious possibilities will suffice. Betz seems to think that this combination of deferral and hedging would, in principle at least, allow for the complete elimination of values from scientific inquiry.

But this strategy may come at a cost. If scientists were able to agree on all reasonable value perspectives with a precise quantification of uncertainty, this strategy might succeed. But that is an ideal case that is unlikely to hold in many policy-relevant scientific domains. Scientists may need to make their description of uncertainty less precise to meet all reasonable evidential thresholds, but as the precision drops, so does the policy-guiding potential of their claims. Scientific conclusions, especially those relevant to policy, should give a relatively clear basis for decision-making. If, for example, climate scientists provided precise probabilities for a variety of future outcomes, policy-makers could incorporate them into decision-making algorithms in order to strategize about climate action. But if climate scientists only provided a set of possible outcomes, with no further indication of their relative likelihoods, it becomes much less clear how to turn that information into policy. While choosing less informative representations of uncertainty helps with deferral and hedging, at some point the representation becomes too uninformative to be useful for guiding policy (in chapter 4 I further articulate this notion of informativeness).

Versions of this objection have been pressed by several philosophers. Steele (2012) notes that in complex domains, deferral would require complicated reports that would decrease the reports' policy-relevance, Elliott (2011) argues that the deferral strategy is harmful because it involves "passing the buck" to decision-makers who must formulate policy, and Steel (2016b) and Brown (2019) claim that the deferral strategy renders science practically inconsequential. Frisch (2020) brings these lines of concern together to argue for a "no-buck-passing" principle for science.

The no-buck-passing principle is a challenge to the end-state desirability of value-freedom. If attaining value-freedom would mean that our scientific conclusions sacrifice their policy-guiding potential, the state of value-freedom would come at a severe cost. This criticism is strong if the appeal of the end-state is a presupposition of one's defense of the value-free ideal, as it appears to be for Betz. However, as will become clear in §3.4, I do not share this presupposition.

To summarize: The version of the value-free ideal offered here is possible to pursue, and the pursuit of this value-free ideal involves routine scientific practices. By satisfying and improving on justificatory principles and practices—the core of the deontic philosophy of science defended in this book—scientists can take steps toward eliminating the influence of values on their inferences, which is characterized as the elimination of bifurcation points. Attending to the heart of science contributes to value-free science, and the active pursuit of value-freedom is, as I now argue, a good thing.

3.4 The Value-Free Ideal Is Pursuit Desirable

Reducing the influence of values on scientific reasoning is desirable, given the benefits of value-freedom discussed in §3.2. Moreover, the elimination of any particular bifurcation point about a hypothesis brings that hypothesis one step closer to being a contribution to common knowledge and hence one step closer to satisfying the constitutive aim of science. However, eliminating bifurcation points can have costs, and the costs trade off against the benefits of value-freedom. The costs may outweigh the benefits so that reduction of the role of values is, all things considered, undesirable. Nevertheless, I argue in this section that despite such undesirability of a value-free end-state, that state is good to pursue.

Before delving into the implications of this possibility, let us first consider these costs. Some may simply be straightforward resource costs. When scientists adopt methodological strategies to reduce value-permeation, the new methods might be costlier in terms of the money, technology, effort, or time required. Improving the quality of our evidence typically implies a greater resource burden. The same problem holds if scientists want to increase the quantity of their evidence to ensure that the degree of justification of a hypothesis does not lie between two contested evidential thresholds. At some point, the marginal cost associated with gathering more or better evidence may outstrip the marginal benefit.

These costs may also be ethical. Conducting a randomized trial in some contexts might help with bias reduction, but it may be ethically unacceptable to administer particular treatments to study participants. Gathering a great deal of evidence on the efficacy of a new drug before publicly presenting the conclusion may strengthen the evidence, but the ethical urgency of a serious disease ravaging a community might mean that we must settle for a smaller quantity of evidence (in chapter 7 I explore the idea that emergencies such as a pandemic can warrant the violation of justificatory norms).

Another kind of cost is the problem with the deferral strategy discussed above—an overly hedged hypothesis might not be able to guide policy. If the aim of a scientific endeavor is to influence policy in particular ways, then scientists must ensure that their conclusions are not hedged to the point that it is unclear how they should inform policy. The purpose of scientific research constrains how the conclusion of that research can be represented.

So, strategies for reducing the influence of values have potential costs in terms of resources, ethics, and policy guidance. When considering whether a particular strategy is desirable, there must be a balance of the costs and benefits. But this cost-benefit analysis is itself an application of normative considerations. If, in attempting to reduce the influence of values, scientists need to make value-based decisions about whether the costs of pursuing value-freedom outweigh the benefits, then do they really move closer to value-freedom? Eliminating a bifurcation point appears to introduce a new bifurcation point corresponding to disagreement about whether eliminating the first bifurcation point was worth the cost.

But recall that I restrict my defense of the value-free ideal to a specific stage of scientific inquiry, the inferential stage, in which scientists draw conclusions from the available evidence. The kinds of value considerations relevant to weighing the costs and benefits of value-minimization strategies are not considerations that apply at that stage. They are, rather, considerations that place prior restrictions on what scientists can accomplish at the inferential stage. This is most obvious when we consider the choice of methodology. Different methodological choices place different constraints on the quality and quantity of evidence from which scientists can draw conclusions. This does not mean that the ethical considerations that go into such methodological choices should be targeted by the value-free ideal. Similarly, a prior specification of the amount of policy guidance required from a particular research project constrains the kinds of conclusions scientists can draw at the inferential stage, but it will not be directly involved as a step in the inferential chain.

The value influence to eliminate when pursuing the value-free ideal is that which makes a difference to the conclusion that is drawn based on evidence—above I referred to this as the difference-to-inference model of value-permeation. The value influence involved in performing the cost-benefit analysis described above plays a different role. That value influence determines the kind and amount of evidence available to scientists (by guiding methodological choices) and the particular form in which the conclusions are presented (for example, how scientists represent uncertainty), but it does not affect the content of the conclusions if the evidence is kept fixed. Disagreements about what level of hedging in scientific results is best for policy purposes or about which method makes the most ethical use of available resources cannot lead scientists to conflicting conclusions from the same evidence base. In other words, these disagreements do not introduce bifurcation points. These decisions should be seen as setting external constraints on what can occur in scientific inference rather than adding considerations internal to inference itself, therefore, such decisions do not threaten the value-free ideal. Conversely, value disagreements that would introduce bifurcation points—disagreements about the appropriate threshold of evidence necessary to draw a particular conclusion, for instance—cannot be assimilated into the external constraints.

Returning to the example of a serious disease ravaging a community, ethical disagreement about the urgency of the situation might lead to disagreement about how much evidence should be collected. A decision about the permissible amount of evidence would limit our ability to eliminate bifurcation points since particular evidence-strengthening strategies, like performing an experiment that would require years to complete, would be ruled out. But the disagreement about urgency is not itself a bifurcation point relative to this set of evidence; it is a disagreement about whether this is all the evidence we can collect. That disagreement may be due to divergent conclusions based on some prior evidence about the virulence of the disease, and in that context, it might constitute a bifurcation point. Ethical considerations involved in bifurcation points in one context of inquiry might function as constraints in another context.

Disagreements about the level of hedging, based on different views about what is required for policy guidance, might lead to bifurcation points. If one group of scientists believes that a more informative conclusion is warranted than another group, then there is an incompatibility in the conclusions drawn. However, the value disagreement leading to this incompatibility is not part of the inferential chain from evidence to conclusion. The disagreement is at a metalevel—once the two groups have determined their conclusions, they are disagreeing about how much they can hedge to eliminate bifurcation points while still fulfilling the policy-guidance constraint. One way to show that the evidence is irrelevant to this disagreement is to hypothetically give both groups the more and less informative conclusions before they even see the evidence. They would still disagree about which conclusion would satisfy the demand for policy guidance. This disagreement would not be about what the evidence reveals; it would be about how the conclusion should be framed (in other words, the disagreement would be about how informative the conclusion should be to sufficiently guide policy—in chapter 4 I discuss the notion of informativeness for scientific assertions).

The Minimal Value State

Imagine a research group or community of researchers trying to reach the value-free end-state by eliminating the influence of values using the strategies discussed above. Their activity, however, is limited by constraints discussed above regarding finite resources, research ethics, and policy guidance. If a value-elimination strategy violates these constraints, they do not use that strategy. As a consequence, the researchers might be unable to attain value-freedom even though they are aiming for it because at some point, there may be no available strategy that takes them closer to the value-free end-state without violating the constraints. In other words, they may reach a stage at which the value-free ideal is no longer pursuit-feasible.

There may be some path dependence involved, that is, how close they can get to the end-state without violating the constraints may depend on what particular sequence of value-elimination strategies they use. In the imagined scenario, the researchers have enough time and persistence to get as close to the end-state as possible across all available paths. Assuming that complete value-freedom is not attainable, the state they ultimately reach will still involve some values making a difference to inference. The state in which the set of values is as small as possible subject to the constraints can be called the *minimal value state*. The minimal value state is the one in which the number of bifurcation points has been minimized while satisfying the constraints.

Even if attaining the value-free end-state is undesirable, arriving at the minimal value state is desirable. This is because the undesirability of the end-state is captured by the constraints on eliminating bifurcation points. All aspects of the end-state that make attaining it not worth the benefits of value-freedom are encoded in the constraints. Since the minimal value state is, by definition, allowed by the constraints, it is a state in which the costs imposed by the constraints do not outweigh the benefits of reducing the influence of values. A properly motivated scientific community should want to arrive at the minimal value state.

This conception of minimality does not call for minimizing *reference* to values in scientific argumentation. It calls for eliminating the extent to which values make a difference to inference, as measured by bifurcation points. This distinction gets to the heart of Longino's concerns about the value-free ideal. She correctly notes that even if scientific reasoning does not explicitly invoke value-based considerations, there is usually still an implicit value framework guiding science. Her call for more diverse perspectives to expose this implicit value system is essentially an imperative to expose the already existing bifurcation points, a stance that I find compelling. Yet once the influence of values is exposed, strategies for value neutralization can be implemented, adjusting the research so that it no longer encodes a particular set of contentious values. Such a strategy involves multiple value perspectives, but it does so strictly as a means to eliminate bifurcation points. As a simple example, one may move from the contentious claims "Based on evidence E, conclusion C_1" and "Based on evidence E, conclusion C_2" to the potentially uncontentious claim "Based on evidence E, conclusion C_1 if you believe in value V_1, and conclusion

C2 if you believe in value V2." This latter conclusion refers to two competing values but may be agreed on by supporters of both. On this account, this diversification of perspectives is therefore a move toward the minimal value state rather than a multiplication of value-based considerations.

The Indeterminacy of Minimality

Attaining the minimal value state is desirable, and pursuing a state of value-freedom within the limits set by the side constraints will, ideally, lead us to the minimal value state. Yet this is not sufficient to establish the pursuit desirability of the value-free ideal.

The pursuit desirability of a goal does not simply amount to the desirability of taking steps toward the goal; it also means that actively and consciously pursuing the goal is desirable. Consider an unattainable and possibly undesirable value-free end-state, S, and suppose the closest we can get to S is the minimal value state M, which is both attainable and desirable. Perhaps by aiming at S, we will end up at M, but that alone does not justify aiming for S. There might, after all, be an even better strategy for getting to M. And such a strategy suggests itself immediately: Rather than aim for S, we could aim directly for M. If we are successful, we will end up in the same place and would have the added advantage of not misrepresenting to ourselves and others our aim. That would suggest abandoning the value-free ideal and replacing it with the minimal value ideal, which would state: Do not aim for an elimination of all values, a target that is both unattainable and undesirable, but, rather, aim for the state of minimal value-permeation.

But while it is true that the minimal value state is our actual desirable end-state, consciously aiming toward that state is generally not an action-guiding strategy for science itself. This is because we have no prior means of determining what the minimal value state is. Its status as the minimal value state only possibly becomes apparent once scientists get there and, attempting to get beyond it, realize there are no further value-elimination strategies they can use without violating the constraints. If a state cannot be recognized as the minimal value state in advance, then we cannot adopt the strategy of aiming for it. The dictum "Attempt to attain a state where your scientific inference relies on a minimum number of bifurcation points" cannot guide action. On the other hand, the value-free state S can be given a prior specification, so one can consciously aim toward it and use it as a guide to further action. It is clear what it means to try to get closer to S. Also, if the process of value-elimination began on a less-than-optimal path and there is path-dependency, then scientists may end up in a local minimum of value influence. Even though there is the possibility of further value-elimination, being in a local minimum would alleviate further motive to eliminate value influence if scientists were aiming at M. So even though the ultimate end regarding value-permeation is the minimal value state, the best action-guiding strategy available to scientists to reach that state is to try to attain complete value-freedom.

Suppose you are trying to find your way in a rocky desert landscape. You want to get to a particular patch of land where a friend buried some treasure years ago. Of course, the treasure is underground, and the patch of ground looks just like any other patch of desert. There are no road signs or addresses, so your friend could not tell you the exact location of the treasure. There is, however, one clearly visible feature—a giant red rock rising in the distance. You do not want to climb the giant red rock; it is a sheer cliff and teeming with snakes. You could not reach the red rock even if you wanted to—it is surrounded by a deep crevasse that cannot be crossed. But you know that the treasure is buried in an accessible area close to the red rock. The best way to reach the treasure is to try to get to the red rock. Once you realize you can get no closer to your pursued goal, the rock, you will be at your actual goal, the location of the treasure. You know that the end-state of being on the red rock is neither attainable nor desirable, but the state you actually want to reach is not one you can practically aim toward. Aiming for the red rock is the best way to get to the place you want to be.

The relationship between the minimal value state and the value-free state is similar. Advances in scientific methods and concepts are often impossible to predict. Thus, it is impossible to predict which values could be eliminated through evidence-strengthening and hedging strategies. As an example, it took statistician Ronald Fisher's methodological innovation of randomization in experimental design to both expose the biasing impact of confounding factors and to design a method to mitigate that biasing impact. Before this development, it was difficult to even conceive of the biasing threat of what is sometimes referred to as *selection bias* or *allocation bias* (Hacking 1988). After this development, experimentalists could block any influence of values that would have intruded via allocation bias. Moreover, the imperative to act as if science should be value-free motivates scientific innovation. Consider the role of publication bias in pharmaceutical research. Critics of this practice have argued that publication bias affords the influence of values (say, industrial executives' profit motive) on conclusions about the effectiveness of pharmaceuticals; in response, clinical science has begun to try to block this threat, for instance, by requiring the preregistration of clinical trials (see chapter 2).

Because the minimal value state is determined by the strategies available for value elimination, and those strategies are often developed in response to immediate scientific challenges rather than predetermined, one cannot know that one is at the minimal value state until one gets there and has run out of ways to get any further toward value-freedom. Like the red rock, the end-state of value-freedom serves as a goal to guide scientific action. Getting to the minimal value state requires researchers to constantly act as if they are trying to get to the value-free state, even if, in moments of reflection and having read the literature on values in science, they may admit that this is not their actual goal. Moreover, the minimal value state can only be recognized based on an inability to get beyond it, and so recognizing that one is at the minimal value state requires active and ongoing attempts to get beyond it, and even then, scientists will be unsure whether further value-elimination is truly impossible or whether a strategy to eliminate values simply has not yet been discovered. So even when they are at what appears to be the minimal value state, scientists should keep trying to go beyond it.

The point is not simply that the minimal value state must be defined or identified with reference to the value-free state. That would not preclude using it as a target to guide scientific action. After all, we often directly pursue aims that are only specifiable in terms of their relation to another putative target. We might aim to order the second-most expensive wine on a menu or park in the space to the left of the closest space to a house. In these examples, however, the reference to another target is needed simply to identify the actual aim, which can then be actively pursued. In the case of the minimal value state, there is no way to actively pursue it independent of the pursuit of value-freedom. As argued above, the minimal value state is neither static nor easily identifiable. Choosing the second-most expensive wine is a well-defined and static goal, and knowing if one has achieved the goal is epistemically trivial. The minimal value state has neither of those properties. Technological developments change the minimal value state, and knowing that one is at the current minimal value state is only possible once one is already there and, crucially, trying to get past it. Thus, the minimal value state cannot be recognized except through the active pursuit of the value-free state because its status as the minimal value state only becomes apparent as we try to get beyond it and find we cannot. Hence, my value-free ideal is conceptualized as *active* rather than *stative*.

It is not just that the minimal value state cannot be identified without aiming at value-freedom. If it were, then one could plausibly characterize the process of bifurcation point elimination as a search for a minimum value state subject to constraints. After all, in many constrained minimization algorithms, the aim is explicitly to find a minimum, but this is accomplished by trying to reduce the relevant quantity as much as possible without violating the constraints. Procedurally, then, a strategy of minimization would look the same as a strategy of elimination.

But this assumes that the minimal value state is static. As argued above, technological and methodological developments in science can change the minimal value state; that is, such developments can permit the elimination of bifurcation points that were ineliminable before the development, all while satisfying the relevant constraints. Thus, the strategy suggested here is not appropriately characterized as straightforward constrained minimization. Besides the exogenous constraints defined by resource limitations, ethical guidelines, and requirements of action-guidance, there is also an endogenous constraint defined by the technological and methodological resources available to the scientist. This is an endogenous constraint because it usually presents as a hard constraint at the level of the individual research project, but when we consider the development of science as a whole, the constraint is malleable.

Science routinely develops technologies that permit the elimination of bifurcation points that previously could not be eliminated. Therefore, while constrained minimization may be an appropriate description of the way scientists should deal with value-permeation at the scale of an individual project, at a larger scale, scientists should be trying to change the minimal value state by changing the technological constraints. Describing the process as one of an elimination of bifurcation points rather than constrained minimization highlights the important lesson that scientific progress does not just involve eliminating bifurcation points that can be eliminated given the current state of technology and methodology. It also involves advancing technology and methodology to permit the elimination of further bifurcation points (in chapter 5, I give a new deontic account of scientific progress that is based on the deontic approach to philosophy of science defended in chapters 1 and 2). There is, in principle, no preidentifiable limit of technological advancement that might permit us to move toward value-freedom. The process of bifurcation point elimination should not be modeled as constrained minimization, nor should it be thought of as aiming toward the minimum value state. It is not just that scientists cannot identify the minimum value state; it is that there is no fixed minimum value state to aim toward.

The indeterminacy of the minimal value state—the fact that we cannot recognize the state except through continuing attempts to get beyond it to value-freedom—is what grounds my argument that value-freedom is pursuit-desirable. It is desirable to consciously aim toward that state, to act as if science should be value-free, even if that state is not a desirable end-state. The pursuit desirability of value-freedom, in turn, grounds my new conceptualization of the value-free ideal.

Larroulet Philippi (2020) makes a related point. Referring to Kitcher's notion of well-ordered science, Larroulet Philippi argues that the notion is an end-state ideal that offers little concrete guidance for how the scientific research agenda should be set. Channeling Sen (2009), Larroulet Philippi notes that end-state ideals are neither necessary nor sufficient for guiding improvements. Sen suggests more emphasis on "transitional accounts," motivating Larroulet Philippi's distinction between "ideal answers" and "ideal procedures." Ideal answers are normative ideals that answer a normative question (like "What does a just society look like?"), while ideal procedures are normative ideals that specify how the normative question should be addressed (like Rawls's "original position"). The value-free ideal has typically been characterized in stative terms, as an ideal answer about the influence of values on science. To put the point another way: The value-free ideal has traditionally been characterized in consequentialist rather than deontic terms. Yet it is impossible to know what values will constitute bifurcation points and what bifurcation points can be eliminated by future methodological developments. Thus the minimal value state for science is indeterminable and therefore not action-guiding. But what science can do is pursue a procedural, active ideal of value-freedom, and the version of the value-free ideal defended here is precisely that: a deontic value-free ideal for science.

3.5 Conclusion

I have argued for a principle that is stronger than merely advising scientists to minimize the influence of values. If my thesis were merely a value-minimization principle, then the view would be similar to that of some prominent critics of the value-free ideal, such as Douglas (2009). Instead, I have argued for a version of the value-free ideal, advising scientists to try to eliminate all value-based influences on scientific reasoning, subject to the external constraints (financial, ethical, action-guidance) that are standard in scientific investigation. In the words of the great physicist Niels Bohr (quoted in Rhodes 1986, 243), science involves "the gradual removal of prejudices." The value-free ideal offered here substantiates Bohr's claim, as the gradual removal of prejudices is precisely what is achieved by the elimination of bifurcation points.

Since I acknowledge that there are value-based constraints on how much value-freedom is desirable, one might say that my view of values in science is not truly a value-free ideal. According to this response, my putative ideal is something like "Scientists should act as if science should be value-free, subject to value-laden constraints." What follows the comma in this statement of the ideal entails that the ideal is not a *value-free* ideal. However, all ideals and goals have this structure. Give an Olympic sprinter the goal of running as fast as she can, and what follows the comma is "subject to value-laden constraints, such as not doping." Give a country the policy of achieving net-zero carbon emissions, and what follows the comma is "subject to value-laden constraints, such as the maintenance of some minimal degree of quality of life." Give a judicial system an ideal that suspects are innocent until proven guilty, and what follows the comma is "subject to value-laden constraints, such as not deciding that all legitimate suspects are innocent." There is nothing inappropriate with saying that the Olympic sprinter has the goal of running as fast as she can, that the country has a policy of achieving net-zero carbon emissions, or that the judicial system has the ideal of suspects being innocent until proven guilty.

The strategies I described in §3.3 to eliminate bifurcation points were all genuinely epistemic, such as evidence strengthening or hedging. Another way to eliminate a bifurcation point could be to exclude or eliminate one of the value sets that constitute a bifurcation point. Consider Aisha and Bheem again, who disagree about a scientific conclusion because Aisha adopts value set V_1 and Bheem adopts value set V_2. So, a bifurcation point exists for Aisha and Bheem. Now suppose that Aisha somehow convinces Bheem to convert from V_2 to V_1. Their bifurcation point would be eliminated. Yet that bifurcation point elimination was achieved not through an epistemic strategy but through a strategy of value-modulation. The value-free ideal here is limited to the elimination of bifurcation points only by legitimate epistemic strategies, namely, those delivered by the Leibniz procedure in service of the end of common knowledge. If we consider Douglas's (2000) example of research on the harmful effects of dioxin, some researchers who prioritized public health values concluded that dioxins are harmful at a particular dose, while other researchers who prioritized commercial values concluded that dioxins are not harmful at that dose. Thus a bifurcation point existed, yet it would be absurd to suggest that removing the bifurcation point could be achieved simply by getting rid of the researchers who prioritized public health values or by convincing these researchers to change their values. Such a strategy to eliminate bifurcation points would clearly not be a justificatory norm delivered by the Leibniz procedure introduced in chapter 1.

Some philosophers maintain that values can play a positive, productive role in scientific reasoning, particularly as suggested by what I refer to as the *value-enrichment critique*. For example, Douglas and Elliott (2022, 203) argue that many well-known cases in the literature show that "values helped to debias science." Douglas (2009, 1) similarly suggests a productive role for values in science: "In place of the value-free ideal, we need a new ideal for science, one that accepts a pervasive role for social and ethical values in scientific reasoning, but one that still protects the integrity of science." Longino (2002, 128) writes that values "can be understood as a rich pool of varied resources, constraints and incentives to help close the gap left by logic." She further suggests that this consideration "turns the value-free ideal upside down—values and interests must be addressed not by elimination or purification strategies, but by more and different values" (2004, 137). To illustrate, Longino (2004, 137) refers to interventions by feminist anthropologists and primatologists since the 1970s as an example of "value-driven research that has improved quality of science in those areas." Brown (2013) goes as far as claiming that evidence should not even have priority over values. In contrast, the view defended here maintains precisely the opposite of these extravagant statements, namely, that scientists should do all they can to resist the influence of values in scientific reasoning.

While I agree that science is often "value-driven" and that value-motivated scientists can improve the quality of research in a particular domain, I suggested above that the cases motivating Longino's position can be reframed in a manner that supports the value-free ideal as I have articulated it here. A more complete description of the cases appealed to by Longino (in various publications cited above), Anderson (2004), and many others is that the infusion of new sets of values helped to debias the relevant science by exposing the impact of preexisting biases, and thereby revealing previously unacknowledged bifurcation points. Though the enrichment of values in a scientific domain can indeed improve the quality of science in that domain, the improvements can be completely characterized in epistemic terms. The enriching values may motivate the pursuit of additional research questions and thereby the gathering of novel evidence, as illustrated by the compelling case described in Anderson (2004), but that is value influence on the external aspects of scientific reasoning. The enriching values may also encourage the search for biases in existing theories and methods, but the ultimate benefit of exposing existing biases is the eventual elimination of newly revealed bifurcation points. The value-motivated exposure of those biases is a crucial step toward eliminating the influence of values on scientific reasoning, like fighting fire with fire.

Nevertheless, one might think that the view defended here is not that different from the views of at least some of the prominent critics of the value-free ideal since these critics agree that value influence should be minimized, not eliminated. I have also argued that the desirable end-state is a state of minimal value influence, not the complete value-free end-state. However, these critics do not explicitly recognize the difference between the pursuit-desirability and end-state-desirability of value-free science. Given my argument that the minimal value state can only be reached by actively pursuing value-freedom, some prominent opponents of the stative characterization of the value-free ideal may support my active characterization of the value-free ideal. That would be welcome.

To summarize, I have argued that it is possible to pursue the value-free ideal, it is good to pursue the value-free ideal, and I have given the value-free ideal a novel active conceptualization: Scientists should act as if science should be value-free. By emphasizing process over product—that is, the desirability of pursuing value-freedom rather than the desirability of attaining it—a deontic philosophy of science provides the foundation for a novel value-free ideal.

CHAPTER FOUR

Scientific Assertion

4.1 Norms of Assertion

Suppose Sasha is a scientist who asserts x in a publication, to policymakers, or in popular media. What norms should we hold Sasha's assertion to? That is, when evaluating Sasha's assertion, what conditions would render that assertion appropriate or inappropriate? Here are a few plausible options:

Sincerity: Sasha must believe x.

Justification: Sasha must have good reasons to believe x.

Truth: x must be true.

Informativeness: x must be sufficiently informative for inference or action.

Combinations of these options are possible:

Knowledge: Sasha must know x.

We could, of course, consider nonepistemic norms, such as whether asserting x will make Sasha rich, but this chapter is concerned only with which epistemic norms apply to scientific assertions. I argue that both *Justification* and *Informativeness* are necessary and together exhaust the epistemic norms for scientific assertions. Despite their popularity in recent philosophical literature, *Sincerity*, *Truth*, and, hence, *Knowledge*, while typically attractive features of assertions, are not necessary for a scientific assertion to be appropriate. Scientific assertions need not be believed or true or known, but they should be justified and informative. This is a deontic account of scientific assertion, and, like the value-free ideal defended in chapter 3, this account of scientific assertion is deontic at two levels: As a norm of assertion, it is itself straightforwardly a deontic principle, and the norm is satisfied when that which is asserted has been justified via the satisfaction of the relevant justificatory norms delivered by the Leibniz procedure described in chapter 1.

By scientific assertion, I mean a claim about the world made from the position of scientific expertise—so if Sasha, a theoretical physicist, is talking about the weather at the office watercooler, she is not making a scientific assertion, but if she publishes an article about string theory, she is making a scientific assertion. A compelling set of norms to assess scientific assertions could be useful. Many scientific assertions are egregiously overconfident and insufficiently justified, and scientific assertions can have profound social and political implications. During the COVID-19 pandemic, we saw how the overconfidence of scientific assertions had profound political implications. I provide examples from the pandemic to illustrate my account of norms of scientific assertion in §4.5.

A norm is a principle, rule, or code that, when violated, entails that the violator should be somehow sanctioned. Norms in this sense are not descriptive but are evaluative and prescriptive—these norms are about proper behavior rather than statistically normal behavior. Moreover, the focus here is on epistemic norms and not on other possible norms that plausibly govern assertion, such as norms of politeness. Williamson (1996) reignited recent philosophical discussion about norms of assertion in general, though before this, Grice's (1989) maxims were norms of assertion, and Grice's concern about "conversational implicatures" is important for scientific assertion.

Like all of us, scientists engage in various sorts of speech acts, including conjectures (Einstein's prediction that the sun will bend light by 1.75 arc seconds), directives (as a graduate student under my supervision, you must learn fundamental statistical theory), commissives (our research group will perform a meta-analysis on the safety of COVID-19 vaccines), and declarations (this book is long). Yet perhaps the most important class of speech acts in science are assertions, or claims about the world. In chapter 1, I argued that the constitutive aim of science is common knowledge, and claims to knowledge are communicated as assertions. Scientific assertions can be expressed in published articles primarily intended to be read by other scientists, in conference presentations, in classroom lectures, in popular media, or in written reports and verbal testimony given to policy-makers.

Since scientific assertions are assertions, one might think that norms of assertion for science are just entailed by norms of assertion in general (in which case, a chapter on norms of assertion in science could be replaced by another epicycle in the literature on norms of assertion in general). Yet, while we can learn much about norms of assertion for science by studying and assessing the existing arguments for norms of assertion in general, the direction of fit ultimately should be reversed: Rather than conforming an account of norms of assertion in science to the best account of norms of assertion in general, an account of norms of assertion in general should be informed by the best account of norms of assertion for science. Most of the literature on norms of assertion is based on intuitions about simple conversational cases. This comprises much of the "linguistic data" (Douven 2006) or "evidence" (Williamson 2000) that is appealed to when judging proposed norms of assertion in epistemology (§4.4). But if we want to investigate epistemic norms of assertion, we should start by considering our best epistemic practice, namely, science, rather than intuitions about simple, contrived conversations about everyday life.

Many of the more plausible putative norms of assertion have been both defended and denied in recent literature. Price (2003), for example, defends *Truth* in general contexts, and there is no reason to think that his position would not apply to science. Price was responding to Rorty (1998), who denied *Truth* (I evaluate Price's argument below). Bird (2022) explicitly follows Williamson (1996), who holds that knowledge is a norm of assertion in general; Bird argues that the aim of science is knowledge, and thus knowledge is a "norm of correctness" for science (though Bird's focus is on the correctness of belief rather than assertion). Gerken (2022) argues for *Justification* for science. Lackey (2007) defends a version of *Justification* while denying that assertions must be believed, thereby rejecting *Sincerity*, and her arguments (based on role responsibilities) are particularly relevant to science. Brown (2020) defends the centrality of sincerity, while John rejects *Sincerity*. Dang and Bright (2021) claim that they reject all of *Sincerity*, *Truth*, and *Justification* for scientific assertion, though we will see in §4.4 that they are in fact committed (and rightly so) to *Justification*. Schroeder (2022) defends something similar to *Informativeness*. So we currently can see a cacophony of conflicting accounts of norms of scientific assertion.

With Gerken and Lackey, and against Dang and Bright, I defend the centrality of *Justification*; though, with Schroeder, I also emphasize the centrality of *Informativeness*. With Dang and Bright, and against Price and many others, I reject *Truth*, and it follows that I reject *Knowledge*. The novelty of this chapter involves defending the importance of *Informativeness*, combining *Justification* and *Informativeness* in a dual norm of assertion, and challenging factive norms for assertion, at least in the context of science. The result is a distinctive account of norms of assertoric discourse in the many forms it can take in science. As suggested above, I believe that the lessons for norms of assertion in science are also lessons for norms of assertion in general, though I keep most of the ensuing discussion within the context of science.

Philosophers of science have not paid much attention to the topic of scientific assertion. In the context of evaluating science, as we saw in chapter 2, philosophers of science have tended to focus on assessing scientific theories or scientific practices. So, broadly construed, when evaluating science, philosophers have tended to focus on the *actions* of scientists. But just as important for the evaluation of science are the *proclamations* of scientists. That is the focus of this chapter.

In §4.2, I give a positive argument for the two epistemic norms that scientific assertions should be held to, namely, *Justification* and *Informativeness*. The centrality of justification for assertion follows from my emphasis on justification in the deontic philosophy of science I proposed in the introduction and further developed in chapter 1. Since factive norms of assertion (such as *Truth* but more usually *Knowledge*) have been attractive to many philosophers, in §4.3, I argue that scientific assertions should not be held to a factive norm. I respond to some potential objections in §4.4 and apply my account to illustrative cases from the COVID-19 pandemic in §4.5. This is a novel account of norms for scientific assertion. While it rejects some existing accounts of norms for assertion, and scientific assertion in particular, it combines some plausible elements of existing accounts to generate a new and distinctive norm (or, rather, pair of norms) for scientific assertions.

4.2 Scientific Assertions Should Be Justified and Informative

Scientific assertions should be both justified and informative. The specific requirements of these norms for particular assertions can change depending on contextual features, such as the consequences of error, the cost of gathering more justificatory resources, domain-specific features of the science in question, and features of one's audience (Goldberg 2015; Gerken

2022; Dethier 2022). In this section, I argue for both a justification norm and an informativeness norm for scientific assertions.

Justification

Scientific assertions should be justified. A first pass at a justification norm of assertion holds:

> Assert *p* only if *p* is justified.

The best and most nuanced description and defense of a justification norm for scientific assertion that I am aware of comes from Gerken (2022), who proposes a "norm of expert scientific testimony." This is a necessary condition for an appropriate scientific assertion based on an adequate degree of scientific justification, where adequacy is relative to a communicative context (2022, 156). Gerken also suggests an additional discursive norm that holds that scientists should articulate aspects of the scientific justification for an assertion. I find this compelling since, as I argued in chapter 1, the constitutive aim of science is common knowledge, and common knowledge requires the public articulation of the justificatory reasons for a scientific claim.

Arguing for the necessity of a justification norm for assertion is straightforward. I do not spend much time arguing for it because it is a fundamental idea held by many writers on the subject, including Williamson (2000), Douven (2006), Lackey (2007), and Kvanvig (2009). The interesting philosophical question is not whether a justification norm is necessary but whether there are, in addition to a justification norm, factive norms such as truth or knowledge. Williamson (2000) and many others claim that there are such factive norms for assertion, though I argue in §4.3 that there is no factive norm of assertion for science.

Most philosophers who argue that there is a factive norm of assertion typically hold that a factive norm is required in addition to a justification norm (see, for example, Price 2003). In a bold article, Dang and Bright (2021) argue against the necessity of justification for appropriate assertion, yet attending to the details of their argument suggests that their favored account of a norm for scientific assertion is perfectly in line with the one offered here, despite what they write (§4.4). So, since virtually all contributors to this topic hold that a justification norm for scientific assertion is minimally necessary, there is little to argue with, though, of course, details of the features of justification itself remain contested (but see Weiner 2005).

We can also consider details of scientific practice that reflect a justification norm of assertion in action. That scientific assertions are held to a justification norm can be understood by observing the many practices in science that are aimed at providing justification, including the domain-specific practices that serve to make science as reliable as possible, such as controlling experiments for potential confounding biases, and the domain-general practices that also aim to make science as reliable as possible, such as critical peer review. The very point of these practices in science is to maximize the extent to which scientific assertions are justified and to criticize those scientific assertions that are not sufficiently justified. As we saw in chapter 1, the core resource in the deontic philosophy of science pursued in this book is the notion of a justificatory norm, and the importance of justificatory norms for science strongly indicates that scientific practice is fundamentally designed to ensure that scientific claims are as justified as possible.

A very basic and fundamental argument for a justification norm of assertion is that we expect people to have good reasons for their assertions. Indeed, this is such a fundamental aspect of assertoric discourse that stating it has the embarrassing quality of stating the obvious. Assertions are not expressions of opinion or preference, and what distinguishes assertions is not merely their world-directed content but also the reasons a speaker has for their assertion. In science, this expectation is even greater. Since justification is the heart of science, scientific assertions must be justified.

The justification of scientific assertions is epistemically accessible by, for example, referring to the methods and results sections of scientific articles. Gerken (2022), channeling Longino (1990), convincingly argues that the justification of a scientific assertion is held to a discursive requirement: Scientific justification requires the provision of reasons that justify an assertion. As noted above, this is a fundamental aspect of scientific practice. In chapter 1, I argued that it is a constitutive requirement of science to not merely share justificatory reasons for scientific claims but to also ensure that these reasons are fully shared, in the sense that a scientific community strives for agreement that a putative justification is indeed justificatory. Many elements of scientific practice are motivated by such a discursive requirement, including peer review, careful descriptions of methods and results in scientific articles, and safeguarding data and sharing it when requested. Violating these practices can result in severe criticism, which reflects a justification norm of assertion being upheld in science.

Informativeness

Science is meant to guide inference and action for individual decision-makers (should I take this drug?), policy-makers (should we implement lockdowns to mitigate the harms of this pandemic?), and other scientists (is this hypothesis worthy of future research?). To achieve those aims, scientists must convey their findings to the public, policy-makers, and other scientists. That involves making scientific assertions. For scientific assertions to guide inference and action, they must be sufficiently informative, particularly in policy-guiding contexts (a point emphasized in chapter 3).

For example, consider a case of excessive hedging, like those from chapters 1 and 3. Suppose Sasha is a climate scientist who has calculated that the global temperature will increase by 2.2 degrees Celsius by the year 2100. Grant that this finding is justified, though, as with all ampliative inferences, it is characterized by some degree of uncertainty. Sasha could assert:

(i) The global temperature will increase by 2.2 degrees Celsius by the year 2100.

The degree of uncertainty could be reduced, and the corresponding degree of justification could be increased, if Sasha hedged her assertion with:

(ii) The global temperature will increase by at least 2 degrees Celsius by the year 2100.

Her justification for her assertion could be increased even further if she asserted:

(iii) The global temperature increase will be greater than 0 sometime in the future.

Her assertion would have maximal justification if she asserted:

(iv) The global temperature will either increase or decrease or not change in the future.

Merely by hedging her assertions, moving sequentially from (i) to (iv), Sasha can increase the justification of her assertions. Yet in doing so, her assertions become less informative. Assertion (i) would be relevant for policy-makers, (ii) a little less so, (iii) would be nearly useless, and (iv) would be entirely useless for guiding policy. This is a problem, as we saw in chapter 3, because a plausible constraint for at least some scientific assertions is that they must be sufficiently informative for guiding policy.

Hedging can, of course, be appropriate in many scientific contexts, as I argued in chapter 3 when considering the impact of values on science. Arguably, scientists should hedge often. However, for scientific assertions to be informative in particular contexts, there is a limit to how hedged an assertion can be (and thus, although there is much in Dethier [2022] that I agree with, I would resist the [unhedged] prescription that "in general scientists should habitually hedge their claims" [2022, 29] because there are contexts in which only unhedged or mostly unhedged assertions could be informative.)

In a compelling article, Schroeder (2022) defends an informativeness norm when scientists present their research to policy-makers. His argument is based on a creative analogy with physicians providing information to patients. Just as physicians are ethically bound to inform patients so they can make good choices about treatment options, scientists also should convey their results in a way that promotes informed decision-making. I believe this view can be generalized to scientific assertions aimed at any kind of audience, including policy-makers but also the public and other scientists since, as above, the public and other scientists also base their inferences and actions on scientific assertions.

For an assertion to be informative, it must be formulated in a way that takes into account features of the intended recipient audience. The intended audience of an article on high theory in physics is typically other physicists. The intended audience of a report describing mitigation strategies for a pandemic is policy-makers. The intended audience of a scientist writing a newspaper editorial is the broader public. Assertions must have particular features to be informative in these differing contexts, a point emphasized by Dethier (2022).

The notion of informativeness relevant here is a formal property of assertions rather than a substantive property. For an assertion to be substantively informative, it must convey information, and if information is factive, then substantive informativeness would be a factive requirement. Thus the informativeness norm would amount to a truth norm. For an assertion to be formally informative, however, it must be articulated in a particular and context-dependent manner such that recipients of the assertion *could* be informed, *were* the assertion true. Consider the following case.

Sasha must decide what to wear today and whether to carry an umbrella. She asks Maria and Dasha about the weather because they have just been outside. Maria says it is 23 degrees Celsius, when it is actually only 14 degrees Celsius, and Dasha says it is possible that it is raining. As a result of Maria's assertion Sasha wears a light blouse, and when she goes outside she is cold, but as a result of Dasha's assertion she has little idea about whether she should bring an umbrella. Maria's assertion is substantively *un*informative but formally informative, as it does not convey true information about the weather, but it is articulated in a precise enough manner for Sasha to make a decision about clothing. Dasha's assertion, however, is substantively informative but formally uninformative, as it conveys true information (albeit a very weak truth) about the weather, but it is articulated in a way that does not help Sasha decide about carrying an umbrella.

What formal informativeness as a norm specifically demands varies from context to context. For example, when reporting the results of a clinical trial, some quantitative ways of analyzing the data are sufficiently informative while others are not (Sprenger and Stegenga 2017; Jäntgen 2023). But in another context, those quantitative methods of data analysis may be irrelevant. For example, when the Intergovernmental Panel on Climate Change forecasts future global climate conditions, the forecasts must be specific enough so that policy-makers can make informed judgments about which policies to enact.

These examples suggest that the notion of informativeness I am working with could be given a formal definition based on asymmetric entailment relations: X is more informative than Y if and only if X entails Y and Y does not entail X (Horn 1972; van Duym 2014). Huber (2008) gives a similar account of informativeness based on probabilities: X is at least as informative as Y if and only if the probability of not-X is at least as high as the probability of not-Y. Floridi (2011) provides a more technical formal definition of informativeness, though his notion of informativeness is factive.

Yet such purely formal accounts of informativeness are incomplete in the context of articulating a norm of scientific assertion because, as mentioned above, informativeness also depends on an interlocutor's interests and capacities. These contextual features can make some assertions too informative. Suppose we are deciding when to stop work for our lunch. You ask me the time, and I say it is thirty seconds after eleven-thirty—my assertion, although very informative unconditional on your interests, contains too much information since you only needed to know that it is not quite lunchtime. Suppose instead that I said it is 1,769,543 milliseconds before noon. Not good at dividing large numbers, you cannot infer what you need to infer. The asymmetric entailment definitions of informativeness stated above hold that my assertion in milliseconds is more informative than if I had just said "eleven-thirty," but that neglects the central function of assertion, namely, to elicit reliable inferences and appropriate actions. Moreover, the entailment account faces the problem of irrelevant conjunction. I could respond to your query about the time by saying, "It is eleven-thirty and London is the capital of Great Britain." That answer entails that it is eleven-thirty, but not vice versa, and so my response is more informative than if I had just said it is eleven-thirty. (Similarly, Grice [1989] considered overinformativeness as something to avoid per his maxim of Quantity.)

One might respond to these considerations by saying that they are pragmatic and not epistemic. Whether an interlocutor has sufficient numeracy, linguistic capability, or sobriety to make an appropriate inference or decision based on an assertion is merely a contingent fact and is extrinsic to epistemic features of the assertion itself. Yet, as Kelp and Simion (2021) argue, an account of assertion can be given in terms of the function of assertion. They claim that the function of assertion is to share knowledge with interlocutors, and they use this to defend a knowledge norm of assertion. One should rather say that the function of assertion is to elicit reliable inferences and utility-maximizing actions in interlocutors. This is not merely pragmatic because reliable inference is a core epistemic concern. Since justification is truth-conducive, the elicitation of reliable inferences in interlocutors is influenced by an assertion's justification, yet the elicitation of reliable inferences in interlocutors is also clearly a function of features of the interlocutors themselves. Thus, I concur with Gerken (2022) and Dethier (2022) that the appropriateness of assertions depends on contextual features, including those of the "common ground" between speaker and interlocutor.

The decision-theoretic notion of "expected value of information" could provide a better formulation of formal informativeness. The general idea is to ask whether an agent would benefit from receiving some claim. The expected value of information asks how valuable learning X would be based on the extent to which learning X would improve one's decisions (Good [1967] is a classic application of this notion in philosophy of science. For a more recent application, see Bradley and Steele 2016). The expected value of information would provide a comparative account of informativeness as follows: X is more informative than Y if and only if the expected utility of acting on the basis of X minus the expected cost of acquiring and utilizing X is greater than the expected utility of acting on the basis of Y minus the expected cost of acquiring and utilizing Y. Let us return to our lunch break example. In response to your question about what time it is, my answer in milliseconds would plausibly be the basis of a worse action than if I had answered "eleven-thirty" (because of your poor numeracy, you cannot use the milliseconds assertion as the basis of a utility-maximizing choice, and using the milliseconds assertion is cognitively costly). This approach is sensitive to the interlocutor features mentioned above. The downside of this approach is that in some contexts it would not provide specific guidance for speakers because, in some contexts, the expected value of one's assertion for a recipient (or, more difficult, a numerous and diverse set of recipients) is not epistemically accessible.

Both justification and informativeness are graded properties, and there is generally a trade-off between the justification of scientific assertions and their informativeness (illustrated by the climate science example above, for instance). This implies that there may be scenarios where scientific assertions cannot be both maximally justified and maximally informative, particularly when resources are constrained and scientists are expected to inform policy-makers as a matter of urgency. This raises many questions about the dual justified and informative norm of scientific assertion. Should both properties be satisfied to some minimum degree for appropriate scientific assertion? Or should the trade-off be permitted to balance anywhere from maximum justification–minimum informativeness to minimum justification–maximum informativeness, depending on context? Should one of the properties get lexical priority? These are interesting and challenging questions, though I suspect there is no context-independent way of answering them. For example, we saw in chapter 3 that nonepistemic values can influence the required justification of a hypothesis via the argument from inductive risk. In chapter 7 I argue that urgent crises can warrant the violation of justificatory norms so that informative policies can be quickly developed. It might also be beneficial for different scientists to balance the trade-off between justification and informativeness in slightly different ways, because this could create dialectically useful debates that foster the articulation, criticism, and improvement of justification. In routine scientific contexts that are not constrained by urgency, however, the deontic approach to philosophy of science—emphasizing justificatory principles and practices as the heart of science—maintains that the trade-off should typically favor justification.

4.3 Against Factive Norms of Assertion

Many philosophers hold that a justification norm of assertion is insufficient, maintaining that assertion is also held to a factive norm such as truth or knowledge (Williamson [2000] is a prominent example; see the references in Schechter [2017] for many more examples). Rosenkranz (2023) describes logical problems with factive notions of assertion. Here I articulate six substantive problems for factive norms of assertion in science, some of which have been raised by critics of factive norms of assertion in general but which have a distinctive edge in the context of science.

Much writing about norms of assertion can seem academic in the pejorative sense, as it involves philosophers trading intuitions back and forth on a seemingly arcane topic. Yet at stake is our judgment about appropriate scientific testimony, particularly in socially relevant science. The example of the violation of the justification and informativeness norms in §4.5 comes from extremely impactful instances of scientific testimony during the COVID-19 pandemic, which serves as a reminder that the stakes for the debate about the norms of scientific assertion are high.

i. No sanctioning justified false scientific assertions

Perhaps the most fundamental argument against a truth norm of assertion is the simple observation that justified false assertions should not be and typically are not criticized. If Sasha has very good reasons that justify x, and Sasha asserts x, we would not sanction Sasha if we later learned that x is false. Thus truth is not a norm of assertion, and neither is knowledge—Douven (2006), Lackey (2007), and Gerken (2011), among others, articulate this criticism. Defenders of factive norms respond that some violations of a knowledge norm of assertion are excusable while others are not—this is the "excuse maneuver" made by, for example, Williamson (2000) and DeRose (2009). Gerken (2017a) and Schechter (2017) criticize the excuse maneuver because it appears ad hoc. The excuse maneuver has spawned a cottage industry of champions and critics, and it is impossible to do justice to the intricacies of the debate here. (In addition to the above references, one could consult Kelp and Simion [2017]; Littlejohn [forthcoming]; Madison [2018]; and Ballarini [2022].) The vibrancy of this debate is enough to suggest that the argument in this subsection will be inconclusive for some readers and decisive for others. Yet if we ask what property would render a speaker blameless when asserting a justified yet false assertion, all should accept this bright and shining answer: the fact that the assertion is justified. A speaker is blameless when asserting a justified yet false assertion precisely because that assertion is justified. If so, the epistemic normativity of assertion lies with justification rather than facticity. We do not criticize speakers for being wrong if they are wrong for the right reasons. The point of a norm is to determine which actions are appropriate and which are sanctionable, and asserting a justified falsehood is not sanctionable; therefore, it is implausible to think of truth or knowledge as a norm of assertion.

Though defenders of the excuse maneuver will remain unmoved, this challenge to a truth norm of assertion is particularly acute in science. This is because justified false assertions are not only not sanctioned in science, justified false assertions are very often *celebrated*. Idealized models, for example, can promote future discoveries and enhance understanding of complex phenomena (Dang and Bright 2021; Potochnik 2017). The Bohr–Sommerfeld atomic model developed in the 1910s was fundamentally incorrect (but see Vickers 2020b), but it was the result of magnificent and creative scientific work. It would be idle to claim that Bohr and Sommerfeld violated a norm by promulgating this model.

That justified false scientific assertions are not appropriate targets of criticism in science contrasts sharply with the fact that *un*justified scientific assertions are routinely criticized. Indeed, unjustified *true* scientific assertions are routinely criticized. For example, an early study suggesting that genes are composed of DNA went against the general assumption of the time that genes are proteins, and scientists expressed many methodological criticisms of the findings (Avery, McLeod, and McCarty 1944). Now, of course, we know that genes are composed of DNA—I discuss this example in more detail in chapter 6. This basic observation about scientific practice is highly suggestive that neither truth nor knowledge are norms of scientific assertion, but justification is.

ii. Science should take epistemic risks

A widely accepted view about science is that scientists should, at least in some contexts, take epistemic risks. Popper, for example, famously argued that scientists should formulate theories that are falsifiable, which involves taking the risk that one's theory could be false. Such activity might not involve assertion per se (Popper's term was *conjecture*). But there are contexts in which scientists are expected to stick their necks out in ways that go beyond mere conjecture, particularly when decisions must be made based on scientific claims (as argued in chapter 3). For example, the results of randomized trials inform regulators about whether a new drug should be approved for clinical use.

More generally, the ampliative nature of scientific inference entails that scientific claims might be false—scientific claims are routinely underdetermined by the available evidence. One might respond that scientific assertions should be hedged to a degree corresponding to the degree of uncertainty of the finding being reported. As mentioned in §4.2, I agree with Dethier (2022) that scientists should more routinely hedge their assertions. In chapter 3, I also defended hedging as a strategy to mitigate the influence of nonepistemic values on scientific reasoning. However, as noted in chapter 3 and earlier in this chapter, there are contexts in which decision-makers need informative, nonhedged assertions. Satisfying a putative factive norm by hedging one's assertion can hamper the informativeness of that assertion. Excessive hedging of scientific assertions can amount to "passing the buck" to decision-makers, rendering the scientific work in question less relevant to decision-makers (Steel 2016b; Brown 2019). We saw in chapter 3 that Frisch (2020) argues for a "no-buck-passing" principle in the context of policy-oriented science. That requires scientists to take epistemic risks. Scientists must occasionally stick their necks out and risk asserting something false. Such assertions are not merely excusable, they can be normatively demanded.

iii. Norm adherence should be epistemically accessible

A driver of a car can know if they are driving under the speed limit. A chess player can know if their move is permitted. A baker can know if they are following a recipe. In these cases, the epistemic accessibility of norm adherence is a good thing. It is generally good when one can determine whether one is following a norm. Norm adherence should be epistemically accessible for both a person governed by a norm and those in a position to evaluate the person's behavior. If one cannot know whether one is following a norm, then that norm loses some of its action-guidance (see [v] below), and that can have bad practical consequences—a driver with no functioning speedometer and no ability to estimate their speed is liable to drive dangerously fast.

In chapter 2 I mentioned the view of norms held by Glüer and Wikforss (2009), who claim that norms should guide action. Norms should stipulate specific behavior when one is in particular circumstances—they have the form "do X when in C." Glüer and Wikforss raise this as an argument against a truth norm for belief, and their argument also works well against a truth norm of assertion. A truth norm of assertion says "assert p when p is true" (there are other versions of a truth norm, but the argument here applies to them all), so a truth norm of assertion includes the truth of the asserted proposition in C. However, the point of science is to discover truths like p. Since we do not know discovered truths until after they are discovered (see next subsection), at the cutting edge of science putative truths cannot be part of C because we do not have sufficient epistemic access to them to be confident of their truth. Given the considerations in the prior subsection, it follows that scientists are often in a position in which they cannot know if they are adhering to a truth norm of assertion. An analogous argument could be formulated against a knowledge norm of assertion.

This feature sets much of science apart from routine epistemic contexts. In such contexts, one often has a pretty good grasp of what is true. I can assert "there is a cup of coffee on my desk," and science fiction scenarios aside, my assertion is both justified and true, and I can ascertain as much, and so I could determine whether I was adhering to a factive norm of assertion. So, epistemic accessibility of adherence to a factive norm of assertion is often not a problem in routine epistemic scenarios. But the inductive and epistemically demanding nature of science entails that scientists often cannot ascertain whether they are adhering to a factive norm of assertion—and that should be taken as a mark not against scientists but against factive norms of assertion.

iv. Truth in science is retrospective benediction

A feature of science that exacerbates the above challenges to a truth norm of assertion is the fact that judgments of truth in science are necessarily retrospective. Before a potential discovery can be deemed true, it must be verified under critical scrutiny, often undergoing iterations of critique and refinement before the discovery is considered certain enough to be deemed true. In chapter 1, we saw that this is a requirement for common knowledge: For a claim to become a contribution to common knowledge, scientists must articulate the putative justification for the claim and convince other scientists that the putative justification is indeed justificatory. Of course, more evidence of different types may be required for the community to become sufficiently confident in the claim's justificatory status and, ultimately, its truth. Moreover, there may be conceptual or ontological barriers to accepting the claim as true, so those barriers would have to be resolved. All that can take time. I call this "retrospective benediction" — the judgment that a putative scientific finding is true involves community-level criticism and uptake, which normally takes time, sometimes years, decades, or even centuries.

An example of how long it can take for a scientific claim to be widely accepted as true is the extremely slow uptake of Copernican theory. Although Copernicus's sun-centered theory of the solar system is often described as a scientific revolution, it took many decades before the theory was widely accepted as true (Westman 2011). A less dramatic but significant example is Watson and Crick's discovery of the double-helical structure of DNA. Their 1953 note in the journal *Nature* used a cautious tone to assert its conclusion, admitting that other models of the structure of DNA were possible. Nine years later, they were awarded the Nobel Prize, by which time the scientific community had generally accepted their conclusion as true. (I explore this argument and the following two arguments in more detail in chapter 5.)

v. Ptolemaic challenge

Just as it can take a long time to prove that a theory is true, it can take a long time to prove that a theory is false. Holding a factive norm for scientific assertion amounts to criticizing scientists who have worked their entire lives within a false paradigm. This is a mark against factive norms of scientific assertion. In chapter 5, I describe in slightly more detail the "Ptolemaic challenge" as an argument against factive accounts of scientific progress, and it is also an argument against factive norms for scientific assertion.

Ptolemaic astronomy was devised in the second century AD by Ptolemy and was based on models of the solar system with Earth at the center (Kuhn 1957). These models could explain a wide range of different kinds of astronomical observations. They were refined over many centuries, and decades after Copernicus proposed a sun-centered model of the solar system in 1543, astronomers were still working with Ptolemaic models. Clearly, all Earth-centered models of the solar system are false. Thus, for nearly fifteen hundred years, the entire discipline of astronomy was spreading falsehoods. Factive norms of assertion sanction the promulgation of falsehoods. Any putative norm of scientific assertion that would call for sanctioning an entire scientific discipline for fifteen hundred years is misguided. Thus, factive norms of scientific assertion are misguided. (A nuance worth noting is the well-known thesis of Duhem [(1908) 1969], which held that Ptolemaic astronomers were not, in fact, proposing theories meant to represent astronomical reality but instead were proposing theories meant only to "save the phenomena," that is, theories that could explain astronomical patterns but were not considered to be true by astronomers. If Duhem is right, then those astronomers were not making assertions about astronomical phenomena, and thus the example could be replaced with one involving genuine assertions.)

Here, too, a defender of a factive norm of assertion could appeal to the excuse maneuver. We saw above that the excuse maneuver is controversial, with some epistemologists relying on it to defend factive norms of assertion and others criticizing it as ad hoc and superfluous. The sheer duration of Ptolemaic astronomy renders the excuse maneuver an implausible response to the Ptolemaic challenge. Excused norm violations should be the exception rather than the rule. If literally every core theoretical assertion for over one thousand years in some domain involves violating a putative norm and every such violation is excused, one could simply and strongly claim that the putative norm is not a norm; instead, the principle that warrants the excusing is the operative norm. And that principle is based on justification.

vi. Truth is a nirvana norm

Imagine someone tells you that you must seek nirvana. What should you do next? It is impossible to say. Perhaps you are already living in such a way that you are approaching nirvana. Or perhaps you are not. With only the imperative to seek nirvana, it is not clear if you should exercise more, which restaurant to dine at for lunch, or what profession you should pursue. For a person to seek nirvana, they need norms that are action-guiding. Buddhism provides such norms in the so-called Noble Eightfold Path. Each of the eight paths includes very specific action-guiding norms. For example, the right conduct path says: do not steal, do not harm others, and do not engage in sexual misconduct; the right mindfulness path says: be conscious of what you are doing, do not be absentminded, and cultivate deep and focused thinking. For a norm to be properly normative, it should guide action. Consider a very basic norm, such as "don't lie to your partner." That norm makes it absolutely clear what (not) to do, and whether you are doing it should be equally clear. A "nirvana norm" is not action-guiding because it is not specific enough in the action it prescribes, and it has the property mentioned above of adherence not being epistemically accessible.

Truth is a nirvana norm for scientific assertion. Grice (1989, 27) articulated his so-called supermaxim as "try to make your contribution one that is true." It is telling that he began with "try to" and immediately added two more concrete action-guiding norms: "do not say what you believe to be false" and "do not say that for which you lack adequate evidence." The effort stipulated in the supermaxim ("try to . . .") is exercised not by what follows "try to" in the supermaxim (". . . make your contribution one that is true") but by maximizing the extent to which one's assertion is justified. (Similarly, as we saw above, Glüer and Wikforss [2009] argue that truth norms are not guiding in the context of belief formation; see Steglich-Petersen [2010] for a response.)

Laudan claimed that truth is a "utopian" norm for science. He famously proposed the "pessimistic meta-induction" as an argument for antirealism, claiming that science cannot attain truth, and even if it could, we would not be able to know it. Many philosophers have criticized Laudan on this; for a recent criticism see Bird (2022). I agree with these critics that truth (or knowledge) is an aim of science, and that science can attain truth (as defended in chapters 1 and 8, respectively). So truth is not a utopian norm. Yet a norm that says "seek truth" or "assert only truths" is a nirvana norm; it is like the imperative that says "seek nirvana" when what is needed is guidance from one of the Noble Eightfold Paths.

4.4 Objections

Truth as Convenient Friction?

Rorty (1998, 19) argued that pragmatists should be "suspicious of the distinction between justification and truth" because the distinction makes no difference to practice, so, for a pragmatist, it should make no difference to philosophical theory. Interlocutors are governed by a norm that requires them to justify their assertions. On this pragmatist view, however, they are not governed by a truth norm because a truth norm makes no difference to assertoric behavior or inquiry. Price (2003, 168) noted that this is an empirical claim, and although testing the claim empirically would be impossible, a thought experiment suggests that when speakers disagree, they "immediately assume" that a truth norm has been breached. Price concluded that truth is "convenient friction" that compels disputants toward resolution — truth is a norm of assertion in general, and there is nothing in his argument to suggest that this norm does not also apply to scientific assertions.

Price (2003, 177) asks us to imagine a linguistic community that views assertions to be mere expressions of opinion, which he calls "merely-opinionated assertions." Abbreviating this term "MOA," Price colorfully calls this community the Mo'ans. The Mo'ans are governed by a sincerity norm and a "personal warranted assertibility" norm (which is basically internalism, or a justificatory standard based on the evidence one has access to), but not a truth norm. Price argues that being governed by only these two norms, the Mo'ans would not be motivated to resolve disagreements. Inconsistent assertions would simply slide past each other like disagreements about tastes or preferences. You like vanilla ice cream, I like chocolate ice cream, and that is that.

Price is aware, of course, that a justificatory norm stronger than mere personal warranted assertibility is available, namely, a community-level standard. But, he asks, what is the relevant community? In the context of scientific assertion, the relevant community might be the current actual community of a scientist. The problem is that current actual scientific communities have imperfect standards, and if we grant that justification can be strengthened by moving from an individual warrantability standard to a current community-level standard, we should also grant that justification can be strengthened by moving from the current community-level standard to a better community-level standard. But why stop there? Indeed, where should we stop? Price (2003, 175) claims, "At each stage, the actual community needs to recognize that it may be wrong by the standards of some broader community." Price concludes that assertion is governed by a norm stronger than a justification norm, namely, a truth norm.

Price's challenge can be met in two steps. The first is to argue that disagreement resolution can be motivated entirely by a community-level justification norm of assertion. The second is to identify the appropriate community-level standard for disagreement resolution.

A justification norm can motivate resolution of disagreement. If two interlocutors are governed by the same community-level justificatory standard and those interlocutors disagree about an assertion, they would each believe that their interlocutor is not justified, and they would each know that their interlocutor thinks that their own position is unjustified. This is sufficient to motivate resolution of the disagreement, particularly in scientific contexts in which assertions are meant to guide inference and action and aid social coordination, especially given that common knowledge is the constitutive aim of science (as I argued in chapter 1). (While the constitutive aim of science is factive, that does not entail that evaluative norms for science must be based on a factive success condition since evaluative norms for science can be deontic rather than consequentialist, as I argued in the introduction.)

Suppose you and I are estimating the number of people who will die from a new virus under a particular mitigation strategy. You estimate and assert x, while I estimate and assert y. Our assertions are meant to guide national policy. Our disagreement motivates a resolution. Yet all the resolving action occurs at the level of evidence and justification: You check the justificatory practice and evidence set that led me to arrive at y, and I check the justificatory practice and evidence set that led you to arrive at x. Since a constitutive aim of science is the development of strong consensus—which requires agreement about justificatory practices—we are motivated to resolve our initial disagreement by investigating our respective justificatory practices.

Disagreement about matters that pertain to social coordination, such as moral or legal principles, provides another antidote to the Mo'an poison. If a disagreement is about matters of taste, then the discourse is governed solely by the norm of sincerity, and no resolution of disagreement is motivated. You like apples, I do not like apples, and that is the end of the matter. Yet moral and political discourse is governed by both a norm of sincerity and a norm of justification, but, plausibly, not a norm of truth since whether there are moral and political truths is, at the very least, controversial, and we have lived with moral and political disagreement for as long as anyone can remember. Such disagreements nonetheless call out for resolution. You think private gun ownership should be impermissible, and I think it should be permissible. It is important to resolve this dispute, regardless of the presence or absence of a truth norm. That is also the case for assertoric disagreement.

The question remains about identifying the relevant community to determine the appropriate justificatory standard, Price's regress argument notwithstanding. I developed an answer in chapter 1. The relevant community is the suitably idealized set of deliberators in the Leibniz procedure. The appropriate community-level justificatory standard for a scientific assertion is the deliverance of the Leibniz procedure. The resulting standard is typically attainable, adherence to the standard is epistemically accessible, and the standard is dynamic, as it improves over time in response to prescriptive guidance.

Both Rorty and Price offered insightful pragmatist accounts of inquiry. Yet I believe both went too far, in opposite directions. Rorty went too far in concluding that truth is unimportant, based on the fact that assertoric discourse is not governed by a truth norm. Price went too far in concluding that because disputants are compelled to resolve disagreements, they must be governed by a truth norm of assertion. Truth is surely important, but it does not follow that assertion is governed by a truth norm, particularly in the context of science.

Justification Not Necessary?

Dang and Bright (2021) describe a case involving physicist William Bragg, whose assertions about radioactivity were not justified based on the total evidence available at the time. These assertions were also false, so Dang and Bright claim that since these assertions were appropriate, for scientific assertion there is neither a justification norm nor a truth norm. If that is correct, then this would be a significant challenge to the account of norms of scientific assertion developed in this chapter. As suggested above, I consider Dang and Bright's arguments against a putative truth norm of assertion to be convincing. And while they claim to be offering an argument against a justification norm of scientific assertion, they describe their preferred "contextualist justificatory account" (8200–1), which strikes me as both compelling and consistent with the justified informative norm of scientific assertion that I offer here.

A few details of the Bragg case are worth mentioning. Dang and Bright (2021, 8193) cite Bragg making some overconfident claims, such as "almost surely" there is "experimental proof" of a material basis of X-rays. Yet many of Dang and Bright's citations to Bragg are quotes of hedged assertions, like the claim that the material basis of X-rays is a "hypothesis" that is "not improbable a priori" (8193) and that should be thought of as a supposition (8192) or a suggestion (8194). A perfectly coherent and correct assessment of Bragg's assertions would be to say that the overconfident assertions were inappropriate and the hedged ones were appropriate (since the hedged ones could be justified). If so, then nothing about the Bragg case would speak against a justification norm of scientific assertion. (And, similarly, while there is much to admire in Fleisher [2021], one could equally resist the claim that "renegade assertions" in research contexts are appropriate and even essential.)

Dethier (2022) argues that the appropriateness of an assertion depends on features of the context in which the assertion is made. What is taken to be common ground influences what is appropriate to assert because it influences what one can legitimately represent oneself as knowing (if one favors a knowledge norm of assertion) or believing (if one favors a sincerity norm of assertion) or considers justified (if one favors a justification norm of assertion).

The specific thesis about a justification norm is difficult to pin down in Dang and Bright (2021), as they simultaneously claim:

(i) apt scientific public avowals need be neither true, known, believed, justified, reasonable to be believed, nor believed to be any such. (8200)

and

(ii) One's avowal must be such that if one's total evidence were what one had gathered in the methodologically proper way for one's latest study, combined with whatever one has taken from the mandated subset of the previous literature contains, then one would be justified in believing one's scientific public avowal. (8199–00)

It is difficult to reconcile these claims. Claim (i) involves denying that scientific assertions must be justified. Claim (ii) entails affirming that scientific assertions must be justified, with a particular characterization of what justification amounts to. So, (i) and (ii) are inconsistent. Claim (i) is aligned with the general message of the article, including its title, but the authors seem to favor claim (ii) in their conclusion, specifically in their "contextualist justificatory account." Upholding claim (i) would require addressing the positive arguments for a justification norm of scientific assertion articulated in §4.2, which Dang and Bright do not address, and which I take to be extremely compelling. Claim (ii), however, is, in broad contours, similar to the justified informative norm of scientific assertion I described in §4.2.

Another general idea urged in Dang and Bright (2021) is that allowing scientists to make "public avowals" that are false and unjustified based on total evidence available can be productive for science. By "public avowal," they mean (a) "utterances made by scientists aimed at informing the wider scientific community of some results," but they do not mean

(b) "statements about the world" (8190). There is a tension here as well. If "public avowal" means (a), then it looks very much like assertion, and the scientific community could be informed by a result only if that result were indeed justified (at least by the evidence generated in that particular scientific work). Yet if "public avowal" does *not* mean (b), then one can say that public avowals are not assertions since assertions are statements about the world. Dang and Bright's general idea is nonetheless compelling and is similar to one of Fleisher's (2021) arguments. It also conveys the essence of Feyerabend's quip that we should let a thousand flowers bloom in science. Yet these thousand flowers can bloom without giving up on the fertilizer of justification by maintaining a distinction between conjecture and assertion. Conjectures are surely important for science — they can motivate research and discovery. Yet conjectures are not assertions. Conjectures might take the same declarative form of assertions, but, following Dethier's insight, the common ground between speaker and interlocutor could indicate that a declaration is a conjecture (say, a hunch worth discussing and pursuing) rather than an assertion (a statement about a putative matter of fact). The same could be said about Fleisher's argument that scientists who advocate and assert theories despite low confidence in them are nonetheless rational; it is surely good for science that scientists advocate and endorse conjectures that may not turn out to be true and may even be poorly substantiated hunches. But when those advocates *assert*, their assertions should be justified.

Linguistic Patterns

Much theorizing about epistemic norms involves philosophers describing quotidian scenarios meant to probe intuitions, which are then used as a consideration in favor of a putative norm. For example, defenders of knowledge norms for belief, assertion, or action describe a scenario in which you have one lottery ticket in a fair lottery for which millions of tickets were sold. Should you believe that you will not win the lottery? Should you say to a friend, "I will not win the lottery"? Should you throw away your lottery ticket? That belief, that assertion, and that action would be odd, even though you have very strong reasons that justify that belief, assertion, and action. What makes them odd? The fact that you do not *know* that you will not win the lottery, claims the defender of knowledge norms. The belief, assertion, and action satisfy a requirement of justification (surely, a belief, assertion, or action that is warranted by a

99.999 percent certainty is strong empirical justification), yet they do not satisfy a requirement of knowledge. Your intuition about the oddness of believing that you will not win the lottery, of saying, "I will not win the lottery," and of throwing away the lottery ticket is explained by the fact that the belief, assertion, and action do not satisfy a knowledge requirement. Thus, goes the argument, belief, assertion, and action are governed by a knowledge norm. Your intuition about the lottery case is "linguistic evidence" for the philosopher concerned with epistemic norms, as Williamson (2000) quaintly describes the results of such thought experiments, and the evidence supposedly supports a philosophical theory: Assertion (and belief and action) is governed by a knowledge norm.

There is nothing wrong with thought experiments—they are a staple of philosophical work. Thought experiments are also crucial in science— Galileo's famous thought experiment about dropping balls from the Leaning Tower of Pisa was very compelling (I return to this example in chapter 6). Indeed, I use thought experiments in several places in this book. Thought experiments help us better understand questions that we might not be able to answer with empirical evidence. Yet, as Papineau notes, commenting on theorizing about "data" from thought experiments: "Many recent epistemologists see their task as developing theories that provide 'the best explanation of all of the intuitive data' (Smithies 2012, 266). Because of this, they are blind to the possibility that these intuitive data are flawed and our epistemological practices defective" (Papineau 2021, 5330). Papineau's cautionary claim is compelling: Much of the "data" in contemporary epistemology is generated by probing intuitions about features of language that are grossly ambiguous. Our everyday linguistic practices and basic intuitions from untutored reflection on the idiosyncrasies of natural language are poor foundations for high theory (see Gerken [2017b] for a detailed argument that we should be cautious about "folk epistemology"). It is scandalous to call the results of such thought experiments "linguistic evidence" in the context of theorizing about epistemic norms (see Levy [forthcoming] for a careful and compelling discussion of the strengths and limitations of thought experiments in philosophy of science).

The scandal appears widespread. For example, in a discussion of whether justification is a gradable property, Hawthorne and Logins (2021, 1848) write, "We can report that ordinary informants tended to classify 'x is justified and y is more justified' as towards the marginal end." But they do not tell us anything about their informants, how many they asked, what "marginal end" means—and they call this "data." For all we know, their informants could have been the people sitting near them in a pub.

Williamson appeals to other putative "linguistic data" to argue that assertion is governed by a knowledge norm. He considers the common English question "How do you know that?" when responding to an assertion to argue that assertion is governed by a knowledge norm because the query presupposes that the speaker knows the assertion. That such a response is perhaps just an accident of English rather than a clue about the deep nature of assertion, or that an equally usual response is "Why do you think so?," does not seem to trouble defenders of a knowledge norm (Mandelkern and Dorst 2022, §4). In the pub near my house, one is less likely to hear such professorial queries in response to an assertion than to hear something rather combative. And such diversity in response to an assertion exists within one idiosyncratic language. To get proper "linguistic evidence" about assertion in the way that Williamson wants would require an assessment of a broader sample of languages.

Consider, for example, a common Ukrainian way of challenging an assertion:

Звідки ти це знаєш?

which is rendered by an online translator just as Williamson would have it: "How do you know this?" Notably, however, the Ukrainian word Звідки is best translated as "from where," which indicates that the recipient of an assertion in Ukrainian using this response is inquiring about the source of the speaker's justification. Another response to an assertion in Ukrainian is:

Де ти це взяв?

This translates as "Where did you get it?" and also indicates a request for a source of justification. Another natural response in Ukrainian is:

Де ти таке почув?

translated as "Where did you hear that?" This question also suggests a request for source of justification. Consider this German response to an assertion:

Warum denkst du das?

which means "Why do you think that?" In Bulgarian:

Откъде знаеш?

Like the first Ukrainian example above, this response to an assertion translates as "From where do you know?" Another option in Bulgarian is:

Кой ти каза?

which means "Who told you (that)?," a response one might also hear in English. Or consider one of many possible examples from Chinese:

为什么你认为 (Wèishénme nǐ rènwéi)

which translates as "Why do you think that?"

One could continue, using examples from any language. Rumfitt (2003) raises a similar challenge in response to Stanley and Williamson's (2001) claim that knowing-how is just another form of knowing-that. Rumfitt argues that if we consider a wider variety of languages, the linguistic arguments made by Stanley and Williamson are equivocal. There are roughly seven thousand languages in use today. The few that I am familiar with use diverse lexical means to express various kinds of justificatory features of assertions, such as modality and information source. The rich variety of ways to convey degrees of uncertainty and reasons for one's assertion are familiar in the Indo-European languages spoken by most philosophers: "I am certain that the structure of DNA is a double helix" directly conveys high confidence, "Nastyusha said there is a good Ukrainian restaurant around the corner" explicitly articulates the source of information, and "The weather forecast says it will probably rain tomorrow" indicates both the source of information and a degree of confidence.

Less familiar to speakers of Indo-European languages is the grammatical feature of evidentiality, which exists in about one-quarter of languages worldwide. An evidential language includes grammatical structures that indicate the kind of information that justifies an assertion (Aikhenvald 2004). Sentences that do not use these features are considered ungrammatical by interlocutors. Evidentiality is not present in most European languages but is common in Indigenous languages of North and South America. For example, in the Tariana language, an Indigenous language of Brazil, a speaker must grammatically mark their information source for an assertion, based on whether the source is visual, nonvisual, inferred specifically, inferred generically, or reported. Simply saying, "José played football" commits a gross grammatical error, but one can say, "José irida di-manika-ka" (the *-ka* suffix means "we saw it"), which translates as "José played football, we saw it") or "José irida di-manika-mahka" ("José played football, we heard it") or "José irida di-manika-nihka" ("José played football, we inferred it from visual evidence") or "José irida di-manika-sika" ("José played football, we assume this on the basis of background knowledge") or "José irida di-manika-pidaka" ("José played football, we were told").

Thus, nonevidential languages have extremely rich ways of articulating the source and degree of justification for assertions, and common linguistic patterns focus on justification in response to assertions, while evidential languages build the source of justification of assertions into the very grammar of language itself. This is highly suggestive. The many ways that uncertainty and the source of information can be conveyed, both lexically and grammatically, suggest that assertion is governed by a justification norm, not a knowledge norm. More modestly, the enriched linguistic evidence conveyed here challenges the weak linguistic evidence that some epistemologists have offered for a knowledge norm of assertion.

4.5 Illustrations

The justification and informativeness norms for scientific assertion can be illustrated by cases in which one or both norms are violated. Two important scientific programs during the COVID-19 pandemic demonstrate the violation of each of these norms. The Bangladesh mask study illustrates the violation of both the informativeness and the justification norm, and the Imperial College London infamous Report 9, based on epidemiological modeling, illustrates the violation of the justification norm.

Bangladesh Mask Study

The Bangladesh mask study was a randomized trial performed from November 2020 to April 2021. Six hundred villages in Bangladesh were part of either an intervention group, which received free masks and information about the importance of masking, or a control group, which received no intervention (Abaluck et al. 2022). The main outcome was symptomatic COVID-19 infection verified with antibody tests. The scale of the study was impressive: 178,322 people were in the mask group, and 163,861 people were in the control group. The main outcome was a clear assertion in the resulting publication: "The intervention reduced symptomatic seroprevalence by 9.5%." The sheer size and methodological quality of this study suggests that this precisely quantified assertion satisfied the justification norm of assertion.

However, very prominent assertions in the publication and subsequent public-facing media did not satisfy either the informativeness norm or the justification norm of assertion. The publication left out the absolute number of subjects in the two groups with symptoms of the virus who tested positive. After commentators asked for this data, the numbers were published: 1,086 cases of symptomatic COVID-19 in the mask group and 1,106 cases in the control group. The reported 9.5 percent reduction in COVID-19 was an effect size known as the "relative risk reduction." I and others have argued that effect size measures like relative risk reduction are not sufficiently informative, at least with respect to theories of decision such as expected utility maximization and theories of inference such as Bayesianism. Effect size measures such as absolute risk reduction, however, can be informative for inference and action (Sprenger and Stegenga 2017; Jäntgen 2023; Kenna and Stegenga 2017; Stegenga 2022a). Yet the Bangladesh mask study publication did not report the absolute risk reduction, which turns out to be a mere 0.08 percent. In other words, a person in the mask group had a probability of getting symptomatic COVID-19 that was 0.08 percent lower than someone in the control group. That is a far cry from the assertion that masks are particularly effective in reducing symptomatic COVID-19.

Three of the study authors then wrote a *New York Times* editorial titled "We Did the Research: Masks Work." This was another clear assertion from scientists speaking as scientists. Yet this assertion is even less informative than the misleading quantitative assertion in the *Science* publication because it broadcasts a claim about the effectiveness of the intervention in generic terms, with no qualification as to how effective the intervention really was in the trial. The *Science* publication also made this generic claim about effectiveness: "Masks are particularly effective in reducing symptomatic seroprevalence of SARS-CoV-2" (Abaluck et al. 2022). What is meant by "particularly effective" is unclear, but on any plausible interpretation of the notion, their results did not support such a claim, particularly given the tiny and more appropriate absolute effect size; therefore, this assertion also violated the justification norm. A similar criticism could be directed at the editorial: On any plausible interpretation of what it means for masks to "work," the data from the trial did not suggest that masks work, in which case the assertion was not justified.

So, in both the *Science* publication and the public-facing editorial, these scientists made prominent assertions that violated both the informativeness norm and the justification norm of scientific assertion.

Report 9

The Imperial College London epidemiology group developed models of the COVID-19 pandemic, which made predictions about the number of deaths and how many critical care hospital beds would be required. They made these predictions under multiple policy scenarios, which included doing nothing and compared doing nothing with some of the lockdown policies that were later enacted (for example, school closing, prohibition of out-of-household social contacts, and so on). This work contributed to the well-known Report 9 given to policy-makers in March 2020 (Ferguson et al. 2020). This report had a huge impact, leading to a shift in the UK government's response to the pandemic and arguably influencing the policy response of other countries (for a good description of the modeling work and its policy context, see Birch [2021]).

The predictions in Report 9 were extremely specific. A visual representation (fig. 2 of Ferguson et al. 2020) displayed very precise predictions under various policy options. Report 9 was extremely informative in the sense described in §4.2: It presented information to policy-makers in a way that was conducive to choosing a course of action.

However, these specific predictions conveyed certainty and precision that were not justified given the limitations of the method. Both epidemiologists and philosophers of science have criticized this modeling work, arguing that the model included very crucial parameters that could have been better empirically warranted and that more sensitivity analyses would have shown that the model's predictions were not robust—see, for example, Winsberg et al. (2020) and van Basshuysen and White (2021) for a critical response. For a subsequent rejoinder, see Winsberg et al. (2021); see also Northcott (2022) for further criticism of the work. In short, Report 9's assertions about the future of the COVID-19 pandemic violated the justification norm of scientific assertion.

4.6 Conclusion

I have defended a norm of scientific assertion that says that scientific assertions must be both justified and sufficiently informative. Factive norms of assertion for science are too demanding in one sense and not demanding enough in another sense. They are too demanding because they require truth. They are not demanding enough because one can assert true claims that are insufficiently informative. Justification is the heart of science, and evaluative norms for science, including the norm of assertion for science, should be based on the heart of science.

CHAPTER FIVE

Scientific Progress

5.1 Introduction

I was surprised to learn that the philosophical literature on scientific progress has neglected a compelling contender. This contender holds that science progresses when there is a change in scientific justification. Justification is central to scientific practice and a pillar of knowledge—hence my surprise.

Understanding scientific progress became important after Kuhn. Kuhn's work was seen as a threat to the rationality of science. If science undergoes revolutions and if a scientific paradigm after a revolution cannot be compared on its epistemic merits to the scientific paradigm before the revolution, then, according to some followers of Kuhn, it is hard to see how science makes progress across scientific revolutions. Scientific revolutions as depicted by Kuhn motivated relativism, skepticism, antirealism, and the nineties science wars.

Though most academics have worked off the hangover of post-Kuhnian extravagance, we now see a widespread public distrust of science. A compelling account of scientific progress could help limit further deterioration of trust in science, at least when such trust is warranted (see chapter 2). An account of scientific progress that is too demanding means that science makes little progress, so it should receive little trust. An account of scientific progress that is not demanding enough means that too many unreliable practices count as scientifically progressive, and so we would place our trust in unreliable practices.

Existing accounts of scientific progress are too demanding or not demanding enough. Many accounts of scientific progress are too demanding because, as I argue in §5.4, they have a truth requirement or something similar: They hold that for science to progress, it must accumulate more truths or more truth-like conclusions. Bird's (2022) "epistemic account" of scientific progress, for example, holds that science progresses when it accumulates knowledge, and knowledge requires truth. Dellsén's (2016) "noetic account" holds that science progresses when scientific understanding increases, and by scientific understanding, Dellsén means the ability to make accurate explanations and predictions, and since accuracy is a factive notion, this account has a truth requirement. What Bird (2022) calls "semantic accounts" are truth-centered: Rowbottom's (2008) account holds that science progresses when it accumulates true scientific beliefs, and Niiniluoto's (2014) verisimilitude account holds that science progresses when it accumulates truths or its theories become more truth-like. The account of scientific progress that I explore here is less demanding, though, as I explain, it is just as demanding as reliable scientific work itself, which is demanding enough.

The reason other accounts of scientific progress are not demanding enough is because they do not require epistemic justification and make scientists seem like tinkerers. Most philosophical discussions of scientific progress seem to assume that if an account of scientific progress dispels with truth, then it must be something like a problem-solving account. Kuhn (1962), Laudan (1977), and more recently Shan (2019) hold that science makes progress as its ability to solve problems increases, and success at solving problems is judged by standards internal to a scientific discipline rather than by a truth standard (for discussion, see Rowbottom [2023]). Yet all sorts of problems can be solved without the progress of science. Defenders of such accounts could say that the problems related to scientific progress are necessarily empirical or theoretical, and thus their solution amounts to scientific progress. But then, progress would occur *because* such solutions would be justificatory.

There is space for an account of scientific progress that sits between the overdemanding truth-centered accounts and the underdemanding problem-solving accounts. My aim in §5.2 is to articulate and defend such an account of scientific progress. The basic idea is to explore a deontic account of scientific progress in which science makes progress when it successfully deploys and develops justificatory principles and practices. I work out some nuances of this account in §5.3. To address what is probably the most obvious question regarding this account, I argue in §5.4 that scientific progress does not require the accumulation of truths or approximation to truths. In short, this is a novel and compelling account of scientific progress with justification, the heart of science, at the center.

Scientific justification is special: It is communal and intersubjective in the sense argued in chapter 1, namely, that justification requires public articulation of justifying reasons, which allows the scientific community to investigate, accept, reject, or revise the putative justification. A complete theory of scientific progress requires that scientific findings have community uptake. Some existing accounts of scientific progress appear to neglect this, though these accounts may implicitly accept that scientific progress is a property of communities. Bird (2022) explicitly and convincingly defends a social account of group belief in science, which could be applied to the notion of scientific progress to uphold a requirement of community uptake. I close the chapter in §5.5 by defending a requirement of community uptake. The justification account of scientific progress is better than existing accounts, as it is consonant with science itself—this is an account of scientific progress faithful to the spirit of the scientific attitude and to the real achievements of science.

5.2 Scientific Progress as Change in Justification

Here is the justification account of scientific progress:

Science makes progress if and only if there is a change in justification.

I fill out some details of this account in §5.3. Here I defend the general plausibility of a justification account of scientific progress. Can an account of scientific progress have justification as its centerpiece?

Almost every philosopher writing about scientific progress seems to think that the answer is no. Rowbottom (2008), for example, suggests that justification is only instrumental for scientific progress. Dellsén (2023) agrees. Though Bird (2008, 280) requires justification, he explicitly states that nothing short of knowledge constitutes progress and that justification, while necessary for knowledge and thus progress, is, without truth, insufficient for progress. I address the requirement of truth for progress more thoroughly in §5.4, where I argue that truth is not required for scientific progress. Here I note that much of science is epistemic in the truest sense of the word; that is, not about truth but about evidence and the evidence-hypothesis relation, aiming to determine what hypotheses or theories are justified based on available evidence and trying to improve those justifications. Scientific practice justly just is about justification.

I address the sufficiency of justification (absent truth or knowledge) as constitutive of scientific progress in §5.4. Dellsén (2016) argues against the necessity of justification for progress. Einstein's 1905 explanation of Brownian motion is a key example. According to the noetic account of scientific progress, the ability of a theory to explain phenomena is a marker of progress; therefore, Einstein's explanation of Brownian motion counted as progress. But this explanation was based on the kinetic theory of heat, which at the time was speculative and so, claims Dellsén, unjustified. Dellsén thus concludes that justification is not necessary for scientific progress. However, precisely because the kinetic theory of heat was able to explain Brownian motion, that theory received some degree of justification since, in general, the ability of a theory to explain a phenomenon provides some justification for that theory. So this is an example of progress, but, contrary to Dellsén's conclusion, this example involved justification.

Dellsén (2023) presents another argument by imagining a scientific discipline with a track record of consistently generating false theories. That dismal track record causes scientists in that discipline to believe that any current theories are probably false, akin to the pessimistic meta-induction for scientific antirealism. Suppose that the discipline develops strong evidence for a theory. This seems like progress, but, claims Dellsén, those scientists would be unjustified in believing that the theory is true (because of the dismal track record in that discipline), and thus, concludes Dellsén, the justification requirement for progress is too demanding. Note, however, that it is the requirement that belief in the *truth* of the theory be justified that is too demanding. This case involves a change in justification for the theory precisely because the case involves the acquisition of confirmatory evidence for the theory. There can be an increase in justification for some hypothesis without there being sufficient grounds to believe that hypothesis because, say, our prior probability in the hypothesis was low, as it would be in the case described by Dellsén. So, if one has the intuition Dellsén has about this case, namely, that it involves progress, the change in justification account of progress accommodates that.

An accumulation of new evidence can increase justification, which would amount to scientific progress. The justification account of scientific progress is more general than the noetic, epistemic, and semantic accounts because it allows for nontheoretical progress. Dellsén (2018, 2), for example, claims that scientific progress is strictly about "improvement in our theories, hypotheses, or other representations of the world, rather than other improvements of or within science." So on this account, accumulation of more evidence would not count as progress, nor would an increase in the confirmation of a hypothesis necessarily count as progress since the mere increase in confirmation of a hypothesis does not need to involve an improvement of understanding about that hypothesis. The noetic, epistemic, and semantic accounts of scientific progress are too theory-centric (Douglas 2014; Shan 2019; and Massimi 2022b make a similar point).

Shan (2019) recently updated the problem-solving account of scientific progress. In this insightful update, the articulation of scientific problems is considered just as progressive as proposing solutions to those problems. I agree that articulation of problems is important. However, without a change in justification, neither the articulation of a scientific problem nor a proposed solution to a scientific problem should be seen as constituting scientific progress. The mere articulation of a scientific problem is like posing a rhetorical question without answering it. Having an articulated scientific problem can be important for the development of some research programs, but it is not necessary. Lucky discoveries can occur without articulated problems, like when Alexander Fleming discovered penicillin. That said, it is plausible to think that having articulated problems can contribute to scientific progress. Yet Bird argues convincingly that contributing to progress does not necessarily constitute progress—just as a large grant for research can promote scientific progress, but it does not itself constitute scientific progress.

Regarding the debate between the view that scientific progress is the accumulation of knowledge and the view that scientific progress is the accumulation of true scientific beliefs, Mizrahi and Buckwalter (2014) tested the intuitions of a large sample of subjects. They found that justification is important for intuitive judgments about what constitutes scientific progress. This finding supports the claim that justification is a necessary component of an account of scientific progress (though the amount of support such results can provide to a theory of scientific progress depends on how strongly one believes in the relevance of untutored folk intuitions for philosophical theories).

One might ask: Justification according to what standard? This account of progress can, in principle, accommodate any view of justification on offer in epistemology, but I believe two options are unattractive. One standard of justification could be strictly internalist, holding that beliefs are justified by the evidence immediately available to an individual scientist holding those beliefs. This would be unsatisfying for an account of scientific progress, as it would render determinations of progress highly individualistic and idiosyncratic (it would also be unsatisfying on independent grounds, as a strictly internalist account of scientific justification is implausible, based on considerations of chapter 1). Another standard might be that of an ideal epistemic community at the end of inquiry. This option would make justification epistemically inaccessible to practicing scientists, thereby losing one of its advantages relative to a truth requirement for progress (§5.4). There is a better account of justification that sits somewhere between these two options. The notion of justificatory norms developed early in this book provides the key. Recall that justificatory norms are those domain-general principles and domain-specific practices in science endorsed by deliberators in the Leibniz procedure (as described in chapter 1), and the use of those justificatory norms renders science more objective (as argued in chapter 2). Scientific justification is based fundamentally on the satisfaction of the justificatory norms delivered by the Leibniz procedure.

5.3 Change in What?

What changes in a change of justification? Here I describe three possible options. The first, based on the number of justified beliefs, is my least favorite. The second and third, based on a notion of graded justification and a notion of change in confirmation, respectively, are both plausible and perhaps interchangeable. A well-developed formal apparatus undergirds the third option, which allows me to explore some nuances, and thus my treatment of the third option is more extensive than the first two.

Number of Justified Beliefs

One way to explicate change in justification, following some recent accounts of scientific progress, is to see it as a change in the number of justified beliefs that science accumulates. Just as Bird argues that scientific progress is the accumulation of knowledge (which entails the accumulation of justified true beliefs), and just as Rowbottom argues that scientific progress is the accumulation of true scientific beliefs, one could hold that a justification-centered account of scientific progress would be based on a change in the number of justified scientific beliefs.

Yet I believe any account of scientific progress based on counting beliefs is implausible. Here is a problem for explicating scientific progress based on the number of beliefs (whether justified or true or both): In any scenario in which there is a justified true belief in x, and further scientific work increases the plausibility of x, a belief-counting approach entails that no progress has been made because there was no increase (or decrease) in the number of beliefs that are justified or true or both (since x was already justified before the additional scientific work, and x is by assumption here true). A plausible example of this is the detection of the Higgs boson in 2012. Before the Large Hadron Collider experiments, the Standard Model of particle physics had a huge amount of empirical and theoretical support. If we consider the hypothesis "the Higgs boson exists," a belief in that hypothesis before 2012 was justified. Yet the Large Hadron Collider experiments that detected the Higgs boson surely must count as scientific progress.

Like the other recent accounts of scientific progress mentioned above, a counting-beliefs approach would adopt an ungraded view of beliefs, which we should be suspicious about. In general, there are good reasons to not hold an ungraded account of belief (such as the lottery paradox). A graded view of doxastic states is also more consistent with scientific practice.

Finally, scientists need not believe claims that they assert as scientifically justified, particularly when many scientists work collaboratively on a project (Dang and Bright 2021). Science is not an institution that simply gathers a set of claims that scientists choose to believe or disbelieve. In general, it is a mistake for philosophers of science to follow epistemologists by developing a belief-centric epistemology of science; rather, scientific epistemology should be centered around the justification of scientific hypotheses or theories, which can be based on credences of individuals but can also involve much more, as we saw in chapter 2.

Change in Degree of Justification

Another way to explicate change in justification is to understand it as a change in the *degree* of justification for a scientific claim. It is very plausible that justification is a gradable property (see, for example, Gerken 2022). Hawthorne and Logins (2021) recently complained about the notion of graded justification, though I find these complaints unconvincing and remain unmoved by them. Justification comes in degrees.

A change in degree of justification can occur when a scientist generates new evidence, when a new hypothesis is introduced that explains existing evidence, or when scientists improve the reliability of methods. A clear instance of a change in degree of justification occurs when newly acquired evidence increases the confirmation of an existing hypothesis. When this happens, science makes progress. Such progress might be modest, or it might be dramatic, as occurred with Arthur Eddington's 1919 observation of the bending of light by gravity, which provided some confirmation of Einstein's general theory of relativity.

The degree of justification of a hypothesis or theory can be influenced by so-called theoretical virtues, such as simplicity, scope, or accuracy, or other nonempirical features. For example, Dawid, Hartmann, and Sprenger (2015) argue that the "no-alternatives argument" can provide some justification for theories. (However, see Okasha [2011] and Stegenga [2015] for a discussion of a problem with appealing to theoretical virtues as a basis for theory choice.)

The example of the detection of the Higgs boson, which was a problem for the belief-counting approach to change in justification, is slightly less problematic for this account. The 2012 detection of the Higgs boson was, of course, evidence for the existence of the Higgs boson and for the Standard Model of particle physics. Yet the existence of both the Higgs boson and the Standard Model were well justified before 2012, and so the work at the Large Hadron Collider could add little justification, which might seem counterintuitive given that detecting the Higgs boson was a truly impressive feat.

One compelling way to articulate a notion of gradable justification is with the tools of confirmation theory, which I turn to now.

Change in Confirmation

One way to explicate change in justification, inspired by recent work in formal epistemology, is to understand it as a change in confirmation.

An obvious case of progress is when new evidence confirms a hypothesis. Yet both increases and decreases in confirmation can constitute scientific progress. A decrease in confirmation can occur in an instance of failed replication: scientists might have a relatively high degree of confirmation for some hypothesis because an initial experiment provided evidence supporting the hypothesis. Yet if a subsequent experiment trying to replicate that initial experiment provides evidence disconfirming that hypothesis, this can count as progress. Indeed, replication failures have become especially important recently due to the so-called replication crisis in psychology. For example, Baumeister et al. (1998) published evidence supporting the existence of "ego depletion," the putative phenomenon whereby subjects' self-control is a limited resource. Larger experiments did not observe ego depletion (Vohs et al. 2020), and such a replication failure should count as scientific progress.

When a scientist introduces a new hypothesis that can explain existing evidence better than available hypotheses, that new hypothesis can undergo a huge increase in confirmation (from zero or undefined to substantial), while the existing hypotheses undergo a decrease in confirmation (Lipton 2004). That is progress. Einstein provided a good example of this in 1915 when his general theory of relativity explained the precession of the perihelion of Mercury's orbit, by then a well-established empirical phenomenon that could not be explained by existing physical theory.

There is a distinction between absolute confirmation, which is the degree to which some hypothesis H is supported by evidence E, and incremental confirmation, which is the extent to which the support of H changes upon getting E. Absolute confirmation is represented by Bayesians as the posterior probability, or the probability of H given E, denoted as $P(H|E)$. There are various measures of incremental confirmation, each with distinct formal representations. Two prominent measures include the difference measure: the difference between the posterior probability and the prior probability, $P(H|E) - P(H)$, and the likelihood ratio measure (or its logarithm), which is the ratio between the likelihood of E given H divided by the likelihood of E given a contrast hypothesis H', $P(E|H) / P(E|H')$ (Fitelson 1999). Since progress implies change, the appropriate notion of confirmation to consider in this account of progress is incremental confirmation. As is standard, $C(H, E)$ represents the incremental confirmation that E provides to H without specifying a particular confirmation measure, and $C_i(H, E)$ represents the incremental confirmation that E provides to H by confirmation measure i.

For subjective Bayesian accounts of confirmation such as that of Sprenger and Hartmann (2019), probabilities are representations of an agent's credences. To address the concern that such a subjective foundation cannot be the basis for characterizing central features of science, Sprenger and Hartmann hold that the agents they are modeling are ideal, rational, and responsive to evidence. Change in confirmation is represented using the formal measures noted above, and the probabilities represent the credence of a rational scientist who responds appropriately to evidence, such that their resulting credence is justified by their evidence.

Whether there is a *unique* way to respond appropriately to evidence is a controversial question in epistemology. For what it is worth, I do not believe there is, yet arguing that point would lead me astray (for differing views on the so-called uniqueness thesis, see White [2005] and Kelly [2014]). If there is a unique way to respond appropriately to evidence, then the formal measures of incremental confirmation are simply representations of that uniquely justified way to respond to evidence. If there is no uniquely justified way to respond to evidence, then there are plausible constraints on justified responses to evidence.

Consider this example: Maria and Sasha want to evaluate hypothesis H, which says "This drug does blah blah." Both Maria and Sasha have the same prior probability in the hypothesis, $P(H)$. A randomized trial is performed that gives evidence (E), suggesting the drug does blah blah. If uniqueness is true—if there is a unique way to respond appropriately to evidence—then when given E, both Maria and Sasha should assign the same degree of confirmation to H. If uniqueness is false, then their assessment of the confirmation provided to H could reasonably differ. Perhaps Maria thinks randomized trials are not as epistemically important as they are often believed to be (having read Worrall [2002]), whereas Sasha thinks randomized trials are more reliable than the alternative (having read Larroulet Philippi [2022]). Yet for both Maria and Sasha, their posterior probability, $P(H|E)$, must be greater than their prior probability, $P(H)$, because E offers at least some confirmation of H (assuming the plausible "positive relevance" definition of evidence). Thus, both Maria and Sasha conclude that H receives some incremental confirmation: For both Maria and Sasha, $C(H, E) > 0$. So, for both Maria and Sasha, according to the confirmation account of scientific progress, this episode involves scientific progress. (When given some other evidence E′, Maria and Sasha might disagree about which of E or E′ provides more confirmation to H and thus disagree about which evidence contributes more scientific progress, but in any case such disagreements are faithful to real scientific disputes.)

One challenge to this approach is that if both increases and decreases in confirmation can count as progress, then there can be a hypothesis that receives first an increase in confirmation and then a decrease of the same amount, and then an increase, and then a decrease, and so on, and this might not seem to amount to progress. Consider confirmation of H by a sequence of experiments 1–N, which generates evidence $E_1, E_2, \ldots E_N$, and we measure confirmation with the difference measure, C_d. We can have:

$$C_d(H, E_1) = x$$
$$C_d(H, E_2) = -x$$
$$C_d(H, E_3) = x$$
$$C_d(H, E_4) = -x \ldots$$

... and so on to N. At the end of this sequence, the posterior probability of the hypothesis would be the same as its prior was before the sequence of experiments began (if N is an even number). It might seem unintuitive to count this as scientific progress, yet it is consistent with the confirmation account of scientific progress.

If an episode in science went through a small number of such iterations, then I would have no problem calling that scientific progress. There are real cases that involve the plausibility of a hypothesis waxing and waning and then waxing and waning again. For example, Margaret Mead (1928) shocked the world with her description of sexually permissive teenagers in Samoa; Derek Freeman (1983) then argued that Mead's evidence was unreliable, and thus the teenage sexual permissiveness hypothesis was disconfirmed; subsequently, Paul Shankman (2009) argued that Freeman had exaggerated his criticisms of Mead and thus the teenage sexual permissiveness hypothesis was plausible, which is roughly where things now stand. However, I doubt that there are many real cases in science that involve more than a handful of such iterations.

Another possible challenge to this approach would be any instance in which there is scientific progress with no change in confirmation. However, I am not aware of plausible examples, and this is arguably because of the analytic relationship between scientific progress and change in confirmation. Consider an example in which you have a prior probability of zero for some hypothesis—that is, you are absolutely certain that this hypothesis is false—and you then acquire evidence about that hypothesis, none of which is confirmatory; there has been an accumulation of evidence but no change in confirmation. Is this progress? I do not think so. That is because hypotheses for which we have prior probabilities of zero are like "Santa Claus exists" or "my body is composed of fewer than seven atoms." Gathering evidence that provides no confirmation for such hypotheses is not scientific progress. The same holds for hypotheses with prior probabilities of one (acquiring evidence that confirms the hypothesis "Santa Claus does not exist" is not scientifically progressive). If a posterior probability of a hypothesis differs from its prior probability—and so there is a change in confirmation—there must be some newly developed confirmatory or disconfirmatory element such as acquisition of evidence for the hypothesis or a refinement of the hypothesis, and such developments are progressive for science.

I have defined scientific progress as a *change* in justification or confirmation. One might be tempted to think that progress should be defined in terms of *increase* rather than merely change. That, however, would have to explain how a replication attempt that does not replicate a finding from an earlier experiment could count as progressive. More substantively, an account of scientific progress in terms of change in confirmation and an account in terms of increase in confirmation are formally equivalent. Suppose we are considering some hypothesis X, and we get evidence E, which provides some incremental *dis*confirmation to X. We can conceive an alternative hypothesis, Y, which says, "not X," and Y, then, gets an increase in confirmation due to E. There is a change in confirmation (specifically, a decrease in confirmation) of X but an increase in confirmation of Y. So "increase in confirmation" and "change in confirmation" are formally interchangeable as an account of scientific progress.

Some increments in confirmation may be minuscule, and sequences of increments in confirmation can involve diminishing returns. Suppose I want to know if a coin is biased to heads. I toss the coin. Heads. I toss again. Heads. I toss again. Tails. Ten tosses, eight heads. One hundred tosses, seventy-seven heads. One thousand tosses, 789 heads. I am now thinking this coin is biased to land heads for roughly 80 percent of coin tosses. The hypothesis "this coin is biased to land heads for 80 percent of coin tosses" received a great deal of confirmation in the first ten tosses, but the amount of incremental confirmation received by the ten tosses between the seven hundredth toss and the seven hundred and tenth toss is much, much less. I raise this point because it will make sense of an important example in §5.4.

* * *

Because the justification account of scientific progress dispels with a truth requirement, it might be seen as closely related to the problem-solving account of scientific progress developed by Laudan (1977) since that account also dispels with a truth requirement. However, Laudan was averse to thinking about scientific progress in terms of justification or confirmation. Whether a theory is "well or poorly confirmed," claimed Laudan (22–23), is irrelevant to assessing progress. All that matters on his account is if a theory can solve a problem. Problems, according to Laudan (25), can be empirical phenomena, and a solution can involve a theory explaining those phenomena, regardless of its confirmation. However, precisely because a theory receives some confirmation when it can explain an empirical phenomenon, Laudan should not have been so resistant to a confirmation account of scientific progress. Yet many instances of changes in confirmation are important and constitute progress but do not contribute to the solution of a problem. The problem-solving account is incomplete, and that is vivid when compared to the justification account.

There is a lot to like about the justification account of scientific progress. It makes sense of routine scientific work, such as generating new evidence. Science progresses with the accumulation of new evidence and not just the refinement of an existing theory or the introduction of a new theory, and so the justification account is more general than theory-centered accounts of scientific progress. It makes sense of the great value of introducing a new hypothesis that explains existing evidence. It makes sense of the importance of experiments aimed at the replication of existing findings, and the interest generated when such attempts fail. In its emphasis on justification, it is relevant to scientific practice. It is given a foundation by our best philosophical theory of scientific confirmation. It entails that scientific progress is epistemically accessible to scientists. What might be seen as its main shortcoming—its lack of reference to truth—is in fact one of its merits, as I now argue.

5.4 Truth as Retrospective Benediction

The justification account of scientific progress dispels with the necessity of the accumulation of truths or related factive notions for scientific progress. Yet a widespread belief is that the aim of science is truth or a related notion, such as knowledge. In chapter 1, I argued that the constitutive aim of science is common knowledge. If the aim of science is truth or knowledge, then it is natural to think that science makes progress as it accumulates truths or knowledge. We saw earlier that several prominent accounts of scientific progress have a truth requirement. My aim in this section is to offer three arguments against a truth requirement for scientific progress. I mentioned these arguments in the introduction and chapter 4, and I develop them here in a little more detail.

Ascriptions of truth are typically retrospective benedictions, a notion we saw earlier. Such benedictions are convenient, as they provide a simple summary of the messy details of scientific work, for allocating credit, teaching students, distributing research funds, and communicating to the public. Truth is retrospective benediction. This is not to say that truth is unimportant—far from it. My point is more modest—to call truth in science a benediction is to emphasize the fact that the ascertainment of truth can take a long time and, obviously and typically, occurs in retrospect.

In real time, scientists are at least sometimes able to ascertain justified changes in confirmation. In real time, scientists are not able to ascertain the achievement of truth. Benedictions of truth take time (Massimi 2016). When Watson and Crick finished building their model of the double-helix structure of DNA, they were confident enough to walk across the street to the Eagle pub in Cambridge to celebrate. They had a clear-eyed assessment of how well confirmed their model was. Yet their one-page 1953 paper in *Nature* was shot through with caution; they claimed that their model was a postulate, based on numerous assumptions, and alternatives to their model, though unlikely, were possible. They were not giving their own finding a benediction. That benediction came nine years later when they were awarded the Nobel Prize. So, in some episodes of scientific progress, benedictions can be made soon after the scientific work itself. In others, however, benedictions can take a very long time. We saw earlier that it took generations of scientists to properly establish Copernican theory (Westman 2011). As Kitcher (1993, 344) puts it, "Scientific debates are resolved through the public articulation and acceptance of a line of reasoning that takes considerable time to emerge."

Laudan (1977) argued that real-time epistemic accessibility of scientific progress is a desideratum for an account of scientific progress—a scientist or a scientific community should be able to ascertain that by doing x, they are making progress. Just as a mountaineer should be able to determine if they are getting nearer to the summit, and just as a baker should be able to determine if the bread is rising, this epistemic accessibility requirement for scientific progress is plausible. Laudan famously argued for antirealism (based on the pessimistic meta-induction); if antirealism is true and one held a truth requirement for scientific progress, then it would follow that science cannot make progress. Laudan took that as an argument against accounts of progress that have a truth requirement. Bird (2022) and others challenge Laudan by directly targeting the argument for antirealism (see my discussion in chapter 8). Yet one can adopt the epistemic accessibility desideratum without adopting antirealism. Here is the general point: The fact that it can take a long time after scientific work occurs for the truth of the findings of that work to receive benediction entails that any account of scientific progress that maintains a truth requirement must violate the epistemic accessibility desideratum.

Similarly, it can take a very long time to learn that one's theories are false and not even approaching the truth. This raises the next problem for maintaining a truth requirement for scientific progress, what I call the *Ptolemaic challenge*. Ptolemaic astronomers toiled for centuries to tally the planets and stars and their positions over time. They developed an Earth-centered model of the solar system based on the geometry of epicycles (a smaller circle placed on the circumference of a larger circle). Their epicyclic models were very successful at explaining their observations, and when they observed anomalous celestial phenomena, they refined their models by adding more epicycles (for a detailed discussion of Ptolemaic astronomy, see Kuhn's 1957 book *The Copernican Revolution*). This research program lasted for many centuries and was based on rigorous observations that were supported by increasingly sophisticated mathematical theorizing. Yet all of those models were false, and over many centuries they were not getting any closer to the truth, as they were all models of the solar system that placed Earth at the center. To maintain a truth requirement for scientific progress requires holding that Ptolemaic astronomy made no progress. *Not a drop.*

Such a view is dismissive of those ancient late-night observers of the starry sky above, those scientists of the oldest science, those curious heirs to Babylon and those diligent students of Aristotle, those scientists who spent centuries in the cold, dark nights of northern Africa recording the movements of stars and planets on clay tablets and devising intricate theories based on models of epicycles on epicycles, those scientists whose forebears designed the pyramids of Egypt to align with the stars and who calculated Earth's circumference to nearly its true value, those scientists who could at least offer a putative explanation of the westward motion of the sky and the eastward motion of the moon relative to the stars and the retrograde motion of planets by layering epicycles on epicycles, and who could predict astronomical observations to within the limits of what could be observed with the naked eye *one thousand years into the future* using epicycles on epicycles, epistemic feats surely more impressive than those that could be achieved today by most lovers of science.

Here we have the Ptolemaic challenge. If an account of scientific progress maintains a truth requirement, it must say that Ptolemaic astronomy made no progress. But Ptolemaic astronomy did make progress. The semantic, epistemic, and noetic accounts of scientific progress face the Ptolemaic challenge. For that reason, they are too demanding.

The Ptolemaic challenge applies to the verisimilitude version of an epistemic account of scientific progress, as there is no sense in which subsequent iterations of Earth-centered models of the solar system were more truth-like. Niiniluoto (2014) distinguishes between real progress and estimated progress, where real progress is based on increasing verisimilitude and estimated progress is based on merely an apparent increase in verisimilitude. He would say that Ptolemaic astronomers merely seemed like they were making progress, but that they were not in fact making progress. This too faces the Ptolemaic challenge.

After centuries of adding epicycles on epicycles, it might have seemed like Ptolemaic astronomy was no longer making progress. Indeed, Lakatos's (1978) attempt to articulate a demarcation criterion for scientific research programs had precisely this sort of concern in mind. Lakatos held that a research program—by which he meant the development and testing of a sequence of theories—is progressive if the sequence of theories makes more and more predictions and if more and more of those predictions turn out to be true. If a research program is not progressive, Lakatos claimed, it is "degenerating." Based on this account, later Ptolemaic astronomy was not progressive; it was degenerating. Yet, as we saw in chapter 2, Lakatos's demarcation criterion placed too much emphasis on novel predictions; while predicted evidence can be more confirming than merely accommodated evidence, that is not always the case, and accommodated evidence can provide some confirmation to a theory (Barnes 2008; Frisch 2015). Moreover, a research program can make little or no progress, but that does not mean it is "degenerating." Ptolemaic astronomy in the late Middle Ages was plausibly in a phase of "diminishing justificatory returns," as mentioned in §5.3—though some incremental confirmation could be gained by adding a thirty-seventh epicycle, it was very little.

Laudan (1984) describes truth as a utopian aim for science because, impressed by the pessimistic meta-induction, he claims that we can never achieve the aim, and even if we could, we could not know if the aim had been achieved. Bird (2022) rightly complains that this is an excessively skeptical position. I agree with Bird that we can often come to know that science has achieved truth (though, as above, in science, that can take a long time). Yet sometimes we cannot know that we have achieved or are approaching truth, and, importantly, sometimes we cannot know that what we now take to be true is in fact false. That was the plight of the Ptolemaic astronomers for many centuries. My position is somewhere between Bird and Laudan. Truth is not a utopian norm, it is a *nirvana norm*. With great diligence some people may achieve nirvana, just as with great diligence science can achieve truth. Yet one might be approaching nirvana and not know it, and, conversely, one might not be approaching nirvana but think otherwise.

As we saw earlier, nirvana norms are not action-guiding and can be used for assessment only retrospectively. Traffic laws guide action—they influence behavior in real time—and of course a police officer can cite a traffic law to justify pulling you over for speeding. A norm that says "seek truth" tells scientists little about how to behave, just as a norm that says "seek nirvana" tells me little about how to behave. Such abstract norms need supplementary, concrete, and action-guiding norms. We saw earlier that in Buddhism, action-guiding norms are expressed in the Noble Eightfold Path. Each path consists of concrete action-guiding norms; for example, the right speech path says: no lying, no rude speech, no idle chitchat; the right livelihood path says: do not earn money by selling weapons, living beings, meat, or alcohol. Telling a person to seek nirvana is to tell them little—it is a nirvana norm. Telling a person to follow the Noble Eightfold Path is to give them very concrete guidance on action. The equivalent concrete norms for science would be whatever justificatory norms are delivered by the Leibniz procedure.

I have given three arguments against maintaining a truth requirement for an account of scientific progress: the epistemic accessibility argument (truth as retrospective benediction), the Ptolemaic challenge, and the truth is a nirvana norm argument. I am not suggesting that truth is unimportant or that science cannot attain truth; rather, I am arguing only that scientific progress is to be judged by reference to changes in justification rather than the achievement of truths or approximations to truth (so, one might consider this a pragmatist theory of progress insofar as some pragmatists dispense with a truth requirement for progress but emphasize justification; see Rorty 1998). Science can discover truths about the world precisely by engaging in its justificatory practices, those justificatory norms delivered by the Leibniz procedure, or by what Nagel (1986) called the "view from nowhere." As I argued in chapter 1, the foundation of scientific justification is a view from nowhere insofar as the ideal deliberators in the Leibniz procedure do not adopt an idiosyncratic view from a particular perspective. However, as also argued in chapter 1, because the constitutive aim of science is common knowledge, scientific justification is fundamentally social. Assessing the progress of science cannot adopt a view from no-who.

5.5 No View from No-Who

Suppose Sasha is searching for the holy grail of science, the ultimate theory of everything. After years of work, she makes a breakthrough discovery, a theory that unifies all physical laws and explains all existing anomalies. She writes up her findings but worries that her discovery may be used to develop terrible weapons. She burns her manuscript, moves to Nepal, and joins a Buddhist monastery, never telling anyone about her discovery and living out her final years in quiet solitude.

Sasha accumulated knowledge, a true finding that could solve many problems and that was, on traditional personalist grounds, justified. Some existing accounts of scientific progress seem to maintain that Sasha made scientific progress—but she did not. I argued in chapter 1 that scientific justification is communal and intersubjective. For a scientific achievement to contribute to scientific progress, there must be not only an in-principle possibility of community uptake but also some actual community uptake (Ross [2020] describes a case like Sasha's to argue that progress should be indexed to the epistemic state of a community rather than individuals). Such community uptake can take time, as occurred with the Copernican model of the solar system, but eventually, such uptake must occur. A finding observed only by the scientist responsible for the finding can hardly be deemed scientific, let alone a contribution to scientific progress. Science cannot make progress with a view from no-who.

The cognitive achievements central to each of the various accounts of scientific progress are idle without community uptake. The existing literature on scientific progress has focused on these cognitive achievements, asking which kind is crucial for scientific progress. Yet Sasha's story shows that this is incomplete.

As Merton ([1942] 1973), Longino (1990), Massimi (2022a), and many others have emphasized, science is a social institution. Aligned with the view developed in chapter 1, many scientists and philosophers of science have held that science is fundamentally public and its methods and evidence must be intersubjectively accessible (argued for by Shapin and Schaffer [1985] and Popper [1959], though for challenges to this publicity requirement, see Goldman [1997]). Moreover, some philosophers argue that scientific communities are themselves epistemic agents (Bird 2022). A scientific hypothesis and its justification must be made public in one way or another, at some time or other, for that hypothesis to contribute to scientific progress. The relevant scientific community must engage with its justification and hold that justification to its standards to determine if a change in confirmation of that hypothesis is indeed justified; if so, the community can do further work on the hypothesis, refining it or using it to discover new findings; if not, further work can be done on it or the hypothesis can be discarded. (Ultimately, the community can conclude whether the hypothesis receives the benediction of truth, but, as I argued in §5.4, progress itself occurs at the moment of justified change in confirmation, not at the moment of benediction.)

One reason in favor of a community uptake requirement for scientific progress is that for future scientific work to develop based on an earlier finding, that finding must be available to other scientists. Another reason is based on Longino's (1990) claim about the importance of criticism in science: If scientific findings are not shared in some way, they cannot be criticized, and criticism is a hallmark of scientific objectivity. Sasha's discovery could not be critically evaluated by anyone else. Still another consideration in favor of a community uptake requirement is Bird's (2022) argument that scientific communities themselves can be the bearers of scientific knowledge. Finally, benediction can occur only if a community uptake requirement is satisfied. (See Bird [2022], chapter 4, which discusses "social knowing." The notion of common knowledge developed in chapter 1 also implies that scientific knowledge is constitutively a community-level state.)

One could say that the work that goes into satisfying a community uptake requirement is not itself epistemic, and it is the epistemic achievement alone that matters for scientific progress. What is subsequently done with that epistemic achievement, such as publication or discussion in the public sphere or education, does not add anything to the epistemic achievement itself. There are many contingent, noncognitive (sociological) reasons that could limit the uptake of scientific findings, but these should not speak against a scientific achievement counting as a contribution to progress. Yet this view is too insensitive to the social structure and function of science. Further, both Harris (2024) and Ross (2020) argue that it is the epistemic states of communities of scientists rather than of individual scientists that matter for scientific progress. (As mentioned above, one could go further and argue that the notion of scientific progress should be based not on doxastic states but on the degree of scientific justification, a nondoxastic notion.)

An interesting example of uptake not occurring can be seen in mathematics today regarding the dispute over whether the so-called abc conjecture has been proven. The abc conjecture is a fundamental conjecture in number theory, and if it were true, many other famous theorems in number theory would follow, such as Fermat's Last Theorem. In 2012, mathematician Shinichi Mochizuki posted on his website a putative proof of the abc conjecture. It was six hundred pages and based on novel mathematical theory that he had developed over years. Mochizuki told no colleagues about his proof, but he had little need to, as rumors were already circulating. Yet when fellow mathematicians began to discuss the preprint, they noted that "it involves ideas which are completely outside the mainstream" and that it was like a "paper from the future, or from outer space" (Ellenberg 2012). Most mathematicians today consider the conjecture still unproven, and some have noted specific flaws in the putative proof. Mochizuki claims that the failure is with other mathematicians and not the proof. I find it plausible to hold that the Mochizuki proof has offered little progress, but if, in the future, mathematicians come to accept the validity of the proof or refinements of it, then progress will be made, but, importantly, the work that went into the proof itself will constitute only part of that progress, as there would also have to be community-level engagement with and uptake of the proof.

5.6 Conclusion

I have offered a new account of scientific progress that is superior to existing accounts. Existing accounts of scientific progress are either too demanding, as with those that have a truth requirement, or are not demanding enough, as with the problem-solving accounts of scientific progress. Science makes progress when there is a change in justification. This account of scientific progress is more in line with scientific practice than competing accounts, as scientific practice is fundamentally centered around practices of justification, and the fallibilism and organized skepticism of science are better suited to a justification-centered account of scientific progress. Finally, an account of scientific progress can only be complete by considering the social structure of science. As in earlier chapters, on scientific evaluation (chapter 2), the value-free ideal (chapter 3), and scientific assertion (chapter 4), relying on the deontic philosophy of science developed in the introduction and chapter 1 affords a novel account of an important evaluative concept in philosophy of science.

CHAPTER SIX

Prise Praise and Prize from Priority

6.1 Introduction

A teacher posing this question to their students hopes for a particular answer: Who dropped objects of different masses from a tall tower, thereby demonstrating that objects of different masses fall at the same speed? The desired answer, however, is very likely incorrect. In 1586, Simon Stevin dropped two lead balls of the same size but different masses from the Nieuwe Kerk in Delft—both balls hit bottom at the same time, thereby disproving a central commitment of the long tradition of physics that followed Aristotle. Galileo's study of the free fall of objects did not begin until three years later, in 1589, and even then, he probably did not actually drop balls, as reputed, from the Leaning Tower of Pisa. Galileo was writing his book *On Motion* during these years, though it was not published until decades later. In that book, Galileo introduced his famous and persuasive thought experiment showing that the assumption of objects of different masses falling at different speeds leads to a contradiction. Galileo is remembered as the scientist who demonstrated that objects of different masses fall at the same speed, but he was not the first to do so. The first may have been Stevin, or perhaps there was another before him. Yet Galileo gets the credit.

Many philosophers who study the allocation of credit in science hold that the so-called priority rule—which states that the first scientist to discover something gets most or all of the credit for that discovery—is the primary principle by which credit in science is allocated. Using technical models, philosophers have noted various consequences of the priority rule. Strevens (2003), for example, following Kitcher (1990), argues that the priority rule promotes an efficient allocation of research resources.

Heesen (2018), in turn, argues that the priority rule promotes hasty, non-reproducible research, Romero (2017) argues that the priority rule discourages attempts to reproduce the work of others (see also Hangel and Schickore 2017), and Rubin and Schneider (2021) argue that the priority rule can cause less-established scientists to be unfairly denied appropriate credit (see also Zollman 2018). Thus, a substantial and growing edifice of scholarly work is predicated on the importance of the priority rule. In this chapter I argue that the priority rule is at best an incomplete account of the reward system in science and I defend an alternative account of scientific reward.

I offer both a descriptive and prescriptive argument. First, it is not generally true that credit is allocated in science according to the priority rule, as suggested by the study of the free fall of objects of different masses. A careful study of examples of credit allocation in science shows that priority is neither necessary nor sufficient to attain credit; credit is often denied to firstcomers but given to secondcomers (those scientists who provide empirical or theoretical support for an existing discovery). Priority is often creditworthy, but not as a general rule. Second, I argue that this is as it should be. The priority rule is not a generally good principle for credit allocation. In addition to the negative consequences of the priority rule noted by Heesen (2018) and Romero (2017) (and long ago by Darwin), there are two fundamental problems with the priority rule: It assumes that scientific achievement is factive, and it assumes that scientific achievement is nonpartible. The facticity aspect of the priority rule holds that credit is and ought to be given to scientists who first discover a truth. The nonpartible aspect of the priority rule holds that credit is and ought to be allocated in a winner-takes-all manner, in which the first scientist to discover something gets all the credit and secondcomers get none. Both aspects of the priority rule are descriptively inaccurate and normatively inappropriate.

I am not concerned with the many contingent, sociological causes of credit allocation in science. The law relating the pressure and volume of gases is called Boyle's law, not Hooke's law or the Hooke-Boyle law, despite the fact that Hooke was Boyle's assistant and an equal collaborator in the discovery of the law, plausibly because when they collaborated, Boyle was rich and Hooke was poor, or perhaps because "Boyle's law" rolls off the tongue more smoothly than "the Hooke-Boyle law." My concern is not with such contingent, sociological causes of allocation of credit but rather with normatively appropriate allocation of credit based on epistemic reasons.

My account of the allocation of credit in science is both descriptively and normatively superior to the priority rule. Building on the deontic philosophy of science developed in the introduction and chapter 1, I argue that the best account of scientific credit is based on the centrality of justification in the epistemology of science. Scientific credit is appropriately awarded to a scientist because she satisfies the justificatory norms delivered by the Leibniz procedure, which may ultimately contribute to common knowledge, and credit can be awarded when scientists develop new justificatory norms or use existing justificatory norms in new ways. I start with a brief background of the priority rule as articulated by Merton (§6.2). I then describe a case from twentieth-century molecular biology in which the priority rule was appropriately flouted, and I add brief descriptions of two other cases which show that the priority rule does not govern credit allocation in science in a fundamental sense (§6.3). Two further problems for the priority rule are that creditworthy science need not discover truth and scientific achievement is partible, but the priority rule assumes otherwise (§6.4). I then give a more general account of credit allocation in science, which is grounded on the deontic philosophy of science developed in chapter 1, and which is based on the degree to which scientific work contributes to changes in confirmation of hypotheses (§6.5), similar to my account of scientific progress in chapter 5. This account of credit allocation is both more descriptively accurate regarding how credit is allocated in science and more normatively appropriate. We should prise praise and prize from priority.

6.2 The Priority Rule

Strevens (2003, 56) describes the priority rule as consisting of two aspects: "First, rewards to scientists are allocated solely on the basis of actual achievement, rather than, for example, on the basis of effort or talent invested. Second, no discovery of a fact or a procedure but the first counts as an actual achievement." So, the priority rule is a winner-takes-all account of scientific credit, in which all credit is allocated to the first scientist who discovers a fact. Rubin and Schneider (2021, 205) seem to agree, claiming "in winner-takes-all cases, none but the first counts as an actual achievement."

Sociologist of science Robert Merton (1957) described the importance of the priority rule in scientific priority disputes. Merton was impressed with the frequency of vitriolic conflicts in science over who made a discovery first, citing many examples. Galileo, in *The Assayer* (1623), challenged his rivals who "attempted to rob me of that glory which was mine." Newton and Hooke battled over priority regarding astronomical findings and optics, and, of course, the Newton–Leibniz controversy regarding the invention of the calculus is famous. Merton wanted to explain the frequency and intensity of these belligerent clashes. He dismissed the idea that being famous in science requires a contentious persona since even very humble scientists such as Cavendish and Watt engage in such disputes. Merton (1957) instead argued that priority disputes are a consequence of "institutional norms of science," most centrally the role responsibility to advance knowledge and the allocation of rewards when that responsibility is fulfilled.

Rewards in science are diverse. Merton's essay noted the importance of eponymy (naming scientific phenomena after the anointed discoverer), which includes such examples as the Planck constant, Halley's Comet, Boolean algebra, Brownian motion, Bayes' theorem, and, recently, the Higgs boson; the naming of units (volt, ohm, coulomb, joule, becquerel, and watt, among many others); the naming of chemical elements (curium, lawrencium, rutherfordium, bohrium, einsteinium, fermium); and, of course, the naming of biological taxa (there are six genera of spiders named after arachnologist Norman Platnick: *Normplatnicka*, *Platnickia*, *Platnickina*, *Platnicknia*, *Platnickopoda*, and *Platnick* [Prendini 2021]). Awards such as the Nobel Prize and the Fields Medal or election to an honorary academic society like the Royal Society are other conduits of credit. Of course, career advancement itself is another form of reward. So, much is at stake for a scientist's work to be deemed original: "In the organised competition to contribute to man's scientific knowledge, the race *is* to the swift, to him who gets there first with his contribution in hand" (Merton 1973, 302). It is this winner-takes-all account of scientific credit allocation that Strevens and other philosophers of science adopt.

The emphasis on priority can have important implications, both positive and negative. Fuller (1988) emphasizes the importance of "essential consensus," which, like the notion of strong consensus introduced in chapter 1, requires scientists to give reasons for their claims and convince their peers that their putative reasons are indeed justificatory. Essential consensus contrasts with "accidental consensus," which arises due to the mere agreement about some claim regardless of the respective reasons for committing to that claim. Fuller argues that competition for priority can have the beneficial result of scientists clearly expressing their justifications for their findings because, if they did not, their colleagues would not give them the appropriate credit. The articulation of putative justification for scientific findings, in turn, entails that any resulting consensus could be essential consensus (or, in my terms, strong consensus) rather than accidental consensus, which ultimately allows for common knowledge (as I argued in chapter 1). Strevens argued for another putative benefit of the priority rule: Using an economic model, he argued that the priority rule contributes to an efficient allocation of research resources, causing the community of scientists to spread research efforts across a diverse landscape of research projects.

Conversely, as noted in the introduction, some philosophers have described negative consequences of the priority rule, particularly insofar as the rule incentivizes speed. This was observed by Darwin, who complained of the "species-mongers" who aim to be the first discoverer of a species and thereby put "a premium on hasty and careless work" (cited in Merton 1973, 300).

Merton's examples show that credit is often given to scientists for being the first to discover something. Such discoveries can be based on either theoretical achievements or empirical achievements—the priority rule simply emphasizes the crucial importance of being the first to accomplish a theoretical or empirical achievement. However, I argue that this is an incomplete account of credit allocation in science. Discovery is one kind of achievement in science, but there are others, and credit is rightly given for all kinds of scientific achievements. Philosophers of science seem to have assumed that the priority rule is *the* principle by which credit is appropriately allocated in science, yet an account of credit based on change in justification is more general, more descriptively accurate, and more appropriate.

6.3 Priority Flouted

Accolades are often given to those scientists who provide empirical or theoretical justification for a hypothesis that has already been formulated and for which there exists some empirical confirmation (secondcomers). Conversely, some firstcomers are denied credit, and while it might seem in such cases that the scientist's achievement has been unfairly neglected, this pattern of credit allocation can be epistemically reasonable.

Consider the discovery that genes are composed of nucleic acids. Until about the middle of the twentieth century, scientists assumed that genes are proteins. In an article published in 1944, Oswald Avery and two colleagues provided the first evidence that genes are composed of nucleic acids. Earlier work showed that it was possible to transform bacterial "types" by heat-killing a virulent type of bacteria and injecting them and another live but nonvirulent type of bacteria into mice; surprisingly, the mice would die, and live virulent bacteria could be extracted from their blood. The live bacteria in the mice had transformed from one type (nonvirulent) to another (virulent), which was possibly some sort of genetic phenomenon. Avery aimed to chemically characterize whatever was causing this transformation, thereby possibly characterizing the basis of heredity.

Avery and his colleagues isolated and preserved the "transforming substance" and performed many chemical tests on it (see Stegenga 2011 and 2013, from which this section draws). Enzymes that degrade proteins did not modulate the capacity of the substance to elicit transformation, but enzymes that degrade nucleic acids eliminated the transformation capacity. The substance had proportions of carbon, hydrogen, nitrogen, and phosphorus that were similar to expected proportions for nucleic acids. Other properties of the substance, such as ultraviolet absorption, electrophoretic movement, and molecular weight were similar to those of nucleic acids. The scientists were aware that they had not perfectly purified the substance and that transformation could have been caused by some other undetected substance rather than nucleic acids. Their article tentatively concluded that the transforming substance was a nucleic acid, such as DNA.

The article was widely cited in the subsequent decade and prompted other scientists to study nucleic acids (Stegenga 2011). Yet during that decade, most scientists did not believe that Avery's work provided much confirmation of the hypothesis that genes are composed of nucleic acids. Some critics noted the concern that their isolation of the transforming substance was imperfect, and thus transformation could be and probably was caused by something other than nucleic acids. Other critics claimed that the phenomenon of transformation did not involve the transmission of hereditary material but rather involved the induction of a genetic mutation. Some believed that nucleic acids were structurally repetitive molecules that lacked sufficient complexity to be the bearers of informative hereditary material. Scientists referred to the findings presented in Avery et al. (1944) in very tentative terms. Prizes awarded to Avery were for his earlier work in immunology. Textbooks continued to describe genes as proteins.

Some of these same scientists who tentatively assessed Avery's findings in the 1940s would, decades later, claim that Avery's discovery was "a total shock and surprise" (Delbrück 1978). What changed?

In the ten years after Avery's results were published, their findings were replicated in different organisms and using different kinds of chemical analyses. Several other findings supported the interpretation that Avery's results were evidence that genes are composed of nucleic acids. For example, Chargaff (1951) demonstrated that the structure of nucleic acid could be sufficiently variable such that nucleic acids could be complex enough to bear genetic information. Others found that bacteria had genes, which was a disputed premise in interpreting Avery's finding as a genetic phenomenon (Lederberg and Tatum 1946). Finally, Alfred Hershey and Martha Chase (1952) used isotopes of radioactive sulfur and phosphorous to selectively "label" proteins and nucleic acids, respectively, in bacteria-infecting viruses. They found that only the phosphorous entered the bacteria, hence only nucleic acids entered the bacteria. The fact that viruses replicate inside the bacteria and that only nucleic acids and not proteins entered the bacteria strongly suggested that nucleic acids and not proteins are the hereditary material. The Hershey–Chase experiment provided compelling evidence that genes are composed of nucleic acids.

A striking and oft-noted fact is that Hershey, along with Max Delbrück and Salvador Luria, was awarded the Nobel Prize (in 1969), but Avery was not (nor was Martha Chase). Yet as suggested in Delbrück's claim quoted earlier, Avery's work was later regarded as providing the first evidence that genes are composed of nucleic acids. That story is enshrined in undergraduate textbooks. Joshua Lederberg, another geneticist working in the years after Avery's famous article, said in his Nobel speech of 1958: "By 1943, Avery and his colleagues had shown that this inherited trait was transmitted from one pneumococcal strain to another by DNA. The general transmission of other traits by the same mechanism can only mean that DNA comprises the genes. To reinforce this conclusion, Hershey and Chase proved that the genetic element of a bacterial virus is also DNA." Given the methodological concerns about the Avery experiment, it was an exaggeration to say that Avery "had shown" that genes are composed of DNA (and certainly not by 1943)—even Avery did not state that so boldly. Yet it would be appropriate to say that Avery was among the first to entertain the hypothesis that genes are composed of nucleic acids and was the first to provide direct evidence for that hypothesis. Hershey and Chase were secondcomers, but their experiment was considered elegant and their evidence compelling. As noted, Hershey and not Avery was awarded the Nobel Prize, and the decade after Avery's 1944 paper saw a subdued reception of the work, while Hershey and Chase's results were immediately celebrated. This is a case of priority being flouted. The firstcomer initially received a little credit and later got more, but never as much as the secondcomer. The simple fact that Hershey was awarded the Nobel Prize is enough to dispel the thought that credit in science is allocated by the priority rule since Hershey was not the first to produce evidence that genes are composed of nucleic acids. If one believes that it was *appropriate* for Hershey to be given such credit (as I do), that should dispel the belief that credit in science *ought* to be allocated according to the priority rule.

One case where the priority rule is appropriately flouted is enough to show that it is not a universal principle guiding the appropriate allocation of credit in science. Yet one might say that warranted exceptions to the priority rule are rare enough, so that it is harmless to consider the priority rule as the single universal principle guiding credit allocation.

Yet I believe cases of flouted priority are common. In the introduction to this chapter I mentioned Galileo as another example of priority being flouted. Here I describe still another example, one in which credit was roughly shared between the firstcomer and secondcomers. In 1896, Henri Becquerel was studying the recently discovered X-rays. He had been investigating uranium phosphorescence and he had a hunch that uranium would absorb sunlight and in turn emit X-rays. He wrapped photographic plates in opaque black paper to block the sunlight, set uranium on top of the wrapped plates, and placed this in sunlight. When he developed the plates, he found outlines of the uranium, and he could create "shadows" of this effect by placing metallic objects like coins between the uranium and the plates. This seemed to confirm his hypothesis. But the weather in Paris that February became cloudy, so he paused the work and put the uranium and photographic plates in a cupboard. A few days later, he retrieved the plates and decided to develop them. He again saw a very clear outline of the uranium, even though no sunlight had hit the uranium in the preceding days. The uranium seemed to have spontaneously emitted radiation. This was the first observation of elemental radiation.

Marie Curie had recently arrived in Paris after leaving her native Poland, as Russian imperialism was attempting to crush Polish culture, and the Russian government had forbidden women from attending universities in its empire. Curie began working on her PhD dissertation at the

Sorbonne in Paris to study uranium radiation, along with her husband Pierre. One of their early findings was that uranium ore was more radioactive than isolated uranium, suggesting that the ore contained other, even more radioactive elements. From the uranium ore they soon isolated radium and polonium, the latter named after Marie's home country (in an irony of history, polonium would later be used by the former imperial conqueror of Curie's homeland in the 2006 murder of Alexander Litvinenko, the Russian spy who defected from Russia and was vocally critical of the Russian government). Radium and polonium were so radioactive they glowed. Curie (1923) wrote, "One of our joys was to go into our workroom at night; we then perceived on all sides the feebly luminous silhouettes of the bottles or capsules containing our products. It was really a lovely sight and one always new to us." She submitted her dissertation in 1903 and that year became the first woman to win a Nobel Prize, for physics, together with her husband and Becquerel. Becquerel may have been the first to discover radioactivity, but the Curies made elements glow brightly as no element had glowed before.

One final example. Marie Curie's daughter, Irène Joliot-Curie, also working with her husband (Frédéric), found in 1934 that radioactivity in some elements could be induced by bombarding them with alpha particles (for this work the couple won the Nobel Prize in Chemistry the following year). Enrico Fermi, also in 1934, replicated this work with neutron bombardment. His experiments bombarding uranium with neutrons were hard to interpret. Fermi thought he had, perhaps, produced a transuranic element (an element heavier than uranium). Although he was wrong about that, he was awarded the Nobel Prize a few years later for his clever empirical work. Chemist Ida Noddack (1934) provided a different interpretation: Fermi had achieved nuclear fission. Physicists generally believed that neutron bombardment could generate elements of higher atomic numbers, yet they did not believe that neutrons were energetic enough to split a heavy atom into two lighter elements, so Noddack's suggestion was difficult to accept. Then Otto Hahn and Fritz Strassmann (1939) provided chemical evidence that neutron bombardment of uranium was generating isotopes of barium, and Lise Meitner and her nephew Otto Frisch demonstrated a theoretical model and then more empirical evidence that what Fermi and Hahn and Strassmann had found was indeed "nuclear fission," which was Frisch's term for the phenomenon of splitting the atom (Meitner and Frisch 1939; Frisch 1939).

Nuclear fission was, of course, a momentous discovery that would profoundly change the world. Who was the first to discover nuclear fission?

Fermi first achieved it but did not recognize it as such. Noddack suggested that Fermi might have achieved nuclear fission, but she did not back this up with sufficient theoretical or empirical warrant to convince her peers. Hahn and Strassmann provided the first empirical grounds to think that Fermi had split the uranium atom, but they did not know how to make sense of this phenomenon. Meitner and Frisch gave the first compelling theoretical explanation of the phenomenon. Hahn won the 1944 Nobel Prize in Chemistry for this work (the award itself was given in 1945 after World War II had ended; nevertheless, the fact that the Nobel committee awarded the prize to a German chemist who had worked on atomic technology for the Nazi regime during the war is extraordinary). Strassman did not win the Nobel Prize, though he was an equal collaborator with Hahn, and Meitner was nominated for the Nobel Prize forty-seven times (Hanel 2015). The magisterial telling of the story by Rhodes (1986) conveys the excitement of the discovery of fission: The Meitner–Frisch papers ignited the American physics community even before *Nature* published them. With just a small touch of historical detail, the question of who deserves credit for the discovery of fission begins to look naive. (For reasons like this, Schindler [2015], referencing Kuhn, argues that it is typically arbitrary to identify a particular scientist at a particular moment as discovering a phenomenon.)

The lesson of the Avery case and others like it can be generalized. Avery received some credit for his preliminary discovery, but Hershey was given more credit. This departure from the priority rule can be put in terms of incremental confirmation (§6.5). Given the methodological concerns about Avery's experiment, his results provided the *first*, but only *some*, confirmation for the hypothesis that genes are composed of nucleic acids. Given the elegance of Hershey's experiment, along with the various shifts in background theoretical context in the years between the two experiments, Hershey's results arguably provided greater incremental confirmation than did Avery's results, even though Hershey and Chase were not the first to provide some confirmation for the hypothesis (in §6.5, I discuss such a credit-by-confirmation account of appropriate allocation of scientific credit). These cases suggest that, rather than being the first to discover a truth that is deemed creditworthy, it is the amount of scientific justification that is deemed creditworthy. The heart of science is justification, and the provision of justification is what warrants the allocation of scientific credit.

A scientific achievement can be described in a variety of ways. Suppose the achievement involves the production of evidence (E_i), which supports a hypothesis (H). The description of the achievement could be the simple claim that "E_i was produced, which confirms H." Or the description could be more detailed by referring to the method (M_i) that generated E_i: "Using M_i, E_i was produced, which confirms H." Now notice that an achievement could be described as the first to accomplish a particular scientific finding on the latter description while not being the first on the former description. Consider the Hershey–Chase experiment. As noted above, this experiment was not the first to produce evidence that genes are composed of nucleic acids—that honor goes to Avery and his colleagues. But if Hershey and Chase's achievement is described not as "production of evidence that suggests that genes are composed of nucleic acids" but rather as "production of evidence that suggests that genes are composed of nucleic acids by radioactively labeling sulfur and phosphorus on bacteriophages and observing which isotope enters bacteria," Hershey and Chase could be considered the first to accomplish this. If so, this case would not serve as a counterexample to the priority rule, and similar descriptions could plausibly be given for all such cases, thereby salvaging the priority rule from the charge of descriptive inadequacy argued for in this section.

Scientific achievements often involve innovations in method. However, when we ask what constitutes a scientific *discovery*, it is plausible to think of an innovation in method as a means to the end of the discovery rather than as part of the discovery itself. The Nobel Foundation's short descriptions of the discoveries of Nobel Prize winners illustrate this. For example, as cited on the Nobel Foundation's website, the Nobel Prize in Medicine and Physiology in 2009 was "for the discovery of how chromosomes are protected by telomeres and the enzyme telomerase," the Nobel Prize in Physics in 2020 was "for the discovery that black hole formation is a robust prediction of the general theory of relativity," and the Nobel Prize in Chemistry in 1911 was for "the discovery of the elements radium and polonium." The descriptions of the discoveries in each case are presented only as ends and do not include whatever means were taken to achieve the ends. That is not to imply that the means—novel experimental methods, for example—are unimportant. Far from it. What facilitates a discovery can rightly be considered a scientific achievement and a contribution to scientific progress, particularly if such facilitation is genuinely justificatory (see chapter 5). A fundamental thesis of this book is that justificatory principles and practices are the heart of science, thereby placing great importance on the scientific means of discovery. The point is that what constitutes a scientific discovery is a conclusion about the world and not what facilitates that discovery. Moreover, as

Strevens argues, there is a symmetry between credit for a scientific achievement and the benefit of that achievement to society. Benefits to society from a scientific discovery usually come from the fact learned through the discovery itself, not from the process of how the discovery was made. For example, Fleming's discovery of penicillin was important to society because it provided a new and powerful antibiotic, and although the story of its discovery is interesting, it does not affect the benefit provided by penicillin. So, the redescription tactic mentioned in the prior paragraph is not a compelling objection to the cases I have described as flouting the priority rule.

6.4 Creditworthiness Is Non-Factive and Partible

The previous section argued that the priority rule is deficient given its emphasis on, well, priority. The priority rule has two further problems. It seems to assume that creditworthy science is both factive and nonpartible, but this is not true. Creditworthy science is non-factive and partible.

Let us begin with the non-factive nature of creditworthy science. An obvious kind of example of creditworthy science involving the "discovery" of a falsehood that is nevertheless considered important and creditworthy is when scientists create a model of a phenomenon that is an idealization of reality. The fact that a model is an idealization implies that it is strictly false. Nevertheless, such models can be important contributions to science (Potochnik 2017); for example, some philosophers argue that false models can be explanatory and provide understanding (Cartwright 1983; Elgin 2017). The sequence of models of atomic structure in the early twentieth century, starting with Thomson's so-called plum pudding model, are good examples. Consider the Rutherford–Bohr model of the atom. The empirical and theoretical work that went into developing this model of the atom with a dense nucleus surrounded by orbiting electrons was impressive and worthy of credit, but it is now seen as a simplification and idealization, at best. Arguably, our current understanding of atomic structure entails that the Rutherford–Bohr model of the atom is straightforwardly false since electrons do not orbit atoms in the discrete electron shells as Bohr conceived (but see Norton [2000] and Vickers [2012] for arguments that Bohr's model contained elements that could later be deemed true according to subsequent quantum theory).

Some creditworthy achievements in science are not even approximations to truth. Consider the example of Ptolemaic models of the solar system discussed in chapter 5. These models could be used to explain virtually all known astronomical phenomena for centuries and to make extremely accurate predictions far into the future. Such work obviously predated the modern institutions of science, and so there was no discernible credit economy of the sort we know today—no Ptolemaic astronomer won the Nobel Prize or appeared on the cover of a glossy magazine as person of the year. But if our contemporary credit economy of science had been in place many centuries ago, Ptolemaic astronomers would certainly have been celebrated, and rightly so. Observing and then offering a putative explanation for planetary retrograde motion, for example, was an impressive feat.

As we saw in §6.2, from the pens of Merton and Strevens the priority rule appears to be an all-or-nothing principle: All credit goes to the firstcomer, and none goes to others. Scientific achievements can, of course, vary in their importance. The value of empirical and theoretical accomplishments awarded by scientific credit are gradable. Think of the work that went into discovering that genes are composed of nucleic acids: Avery surely deserved some credit, as his work provided the first empirical evidence that genes are composed of nucleic acids; Hershey surely deserved some credit, as his work provided very compelling evidence that genes are composed of nucleic acids. The very fact that we can ask how much credit each one deserved suggests that credit itself is gradable. An account of credit allocation in science should be based on the fact that the value of scientific achievements is gradable. (Some recent work in the formal modeling of credit allocation grants that credit is gradable, such as Heesen [2019].)

Moreover, credit for a particular scientific achievement is partible, so our conception of scientific credit should allow for the sharing of credit. The priority rule, as we saw in §6.2, is a winner-takes-all account. Emphasizing this, Strevens (2011, 192) claims that the first group of scientists to make a discovery "get all the credit" while a second group "just for being late, get nothing." So, according to the priority rule, scientific credit is nonpartible, yet the justificatory work contributing to a scientific discovery is routinely distributed across many scientists, as we saw with the example of uranium fission. There are, of course, many cases of scientific credit being shared, as when several scientists contribute to different aspects of a particular scientific discovery. For example, Watson and Crick developed their models of the structure of DNA by using the x-ray crystallography work of Franklin and Wilkins, and the resulting Nobel Prize was awarded to members of both groups (Franklin had died by the time the prize was awarded in 1962, but the other three received the prize). A nonpartible account of scientific credit has no problem with such cases because one could simply say that it was the overall set of collaborating scientists that was the firstcomer and thus the credit should be shared among the set.

More difficult for nonpartible accounts of scientific credit are cases in which independent scientists (or groups of scientists) each contribute evidence that supports a discovery, and credit for the discovery should be at least partially shared. As we have seen, the chemical characterization of the gene is such a case. Avery surely deserved some credit for discovering that genes are DNA. While I believe it is an exaggeration, many scientists claimed that Avery should have been awarded a Nobel Prize; in any case, Avery's 1944 article was cited hundreds of times and has been canonized in textbooks. Just as surely, Hershey deserved (and received) credit for the same discovery. The existence of cases in which credit is appropriately divided between independent groups of scientists, particularly when the contributions of those scientists are independent of each other, contradicts the view that credit in science is nonpartible and that all credit goes to the firstcomer.

6.5 Credit by Confirmation

The emphasis of the priority rule on being the first to discover a putative fact makes science seem like a race. Of course, sometimes science involves races, exemplified by Watson and Crick's race to discover the structure of DNA before their competitors did. Yet as a general metaphor to consider the appropriate allocation of credit in science, there are other possibilities. A world-class violinist is not praised for being the first to do something among competitors, such as in a race, but rather for being among the most technically skilled. A judge is praised not for giving a fast verdict but for giving a verdict that is just and fair.

The examples in §6.3 are suggestive: The allocation of credit in science is generally a function not of priority but (often, and ought to be) of the degree of justification that some particular scientific work generates. The examples also suggest that there are various kinds of achievements in science that can be creditworthy, including the production of compelling evidence for a theory, the recognition of some evidence as resulting from a particular kind of phenomenon, and the articulation of a theory to explain existing phenomena.

In §6.3, I argued that we can describe a discovery by referring to what is learned about the world from the discovery, without referring to the justification for that discovery. Of course, the justification is crucially important. We can think of the relationship between discovery and justification as one of end and means. Justificatory practices are means to the end of discovery. We can then ask: Should credit be assigned for successfully attaining the end, or should credit be assigned for appropriately deploying means? These questions are distinct because appropriately deploying means does not guarantee that the end will be attained, and successfully attaining the end does not universally require the appropriate deployment of means.

A consequentialist about scientific credit would say that we should assign credit for successfully attaining the end or aim of science. In chapter 1, I argued that the constitutive aim of science is common knowledge, so one plausible end-oriented principle of credit allocation would be to assign scientific credit for contributions to common knowledge. Even with this credit allocation principle, there would be an emphasis on justification, as justification is central to common knowledge. A nonconsequentialist about scientific credit, however, would say that we should assign credit for appropriately deploying means. I gave reasons in the introduction, illustrated with the violinist example, to be a nonconsequentialist about goal-seeking in general, and such reasons apply to credit allocation. The deontic philosophy of science developed earlier holds that our evaluation of science should be indexed to the satisfaction of justificatory norms, and perhaps to the development of new justificatory norms, and this applies to the allocation of scientific credit.

As discussed in chapter 5, the notion of degree of justification can be grounded by reference to confirmation. We saw that philosophers distinguish between final or absolute confirmation from incremental confirmation, with some attempting to defend one particular measure of incremental confirmation as the best (Fitelson 1999). Yet arguments exist for multiple confirmation measures, which suggests that there might not be a single formal measure that adequately represents all instances of confirmation or change in confirmation. I propose that the appropriate allocation of credit in science is a function of the extent to which scientific work contributes to confirmation or change in confirmation, as represented by any of the leading measures of confirmation. That this credit allocation principle is distinct from the priority rule should be obvious because, as we saw with the examples in §6.3, secondcomers can contribute a great amount of confirmation. That this credit allocation principle can nonetheless account for the fact that many firstcomers are appropriately awarded credit, examples of which we saw in §6.2, should be equally obvious: Often, being the first to provide empirical evidence or theoretical explanation can involve a great amount of confirmation.

Some creditworthy scientific achievements might be adequately represented as conferring a large final confirmation. The detection of the Higgs boson in 2012 arguably did not add much confirmation to the hypothesis that the Higgs boson exists since it made sense to have a very high prior probability on that hypothesis before the detection, based on the many empirical successes of the Standard Model of particle physics (as discussed in chapter 5). If so, then the detection of the Higgs boson would have small values for any of the measures of incremental confirmation. But the value for the final confirmation was arguably very high.

Other creditworthy scientific achievements might be adequately represented as conferring credit according to one of the measures of incremental confirmation. Avery's results, for example, did not confer a large or even middling posterior probability of the hypothesis that genes are composed of nucleic acids, so it could not be said to have provided much confirmation according to the difference measure. However, since the prior probability of that hypothesis was at the time very low, Avery's evidence could be said to have provided a large degree of confirmation according to the ratio measure. Hershey's evidence arguably had the converse features.

Though this account of scientific credit departs from Strevens's (2003) account of the priority rule, it is closer to his subsequent account of credit, which holds that scientific reward is proportional to the actual contribution to society, and actual contribution is proportional to "epistemic security," or justification (Strevens 2006).

One might be concerned that the various confirmation measures give significantly different values of incremental confirmation for the same scientific work. An account of appropriate credit allocation in science based on this plurality of confirmation measures entails that there is no unique metric of scientific achievement and thus no unique amount of credit that ought to be allocated to a particular scientific achievement. Floating down this river of ambiguity and uncertainty that we call life, this might trouble some readers. Yet this aspect of my account of scientific credit is a feature, not a bug. Scientists often agree about the relevant facts of a scientific achievement but disagree about their importance.

Disagreements about the amount of credit deemed appropriate for a particular scientific achievement can be based (implicitly) on differing views about which confirmation measure is appropriate for gauging achievement. Disagreements about how much credit Avery deserved for his work on the chemical composition of genes, for example, illustrates this.

6.6 Conclusion

Science does not generally allocate credit according to the priority rule. Nor should it. Rather, credit is often and ought to be allocated on the basis of achievements of various types, which can be represented by the various measures of confirmation.

The argument in this chapter illustrates the guiding thesis developed in the early chapters of this book, namely, that our evaluative concepts for science should be articulated with reference to justification, the heart of science. The priority rule emphasizes speed and facticity, while appropriate credit allocation emphasizes contributions to scientific justification. Sometimes, though, speed in science is important, an idea I turn to in chapter 7.

CHAPTER SEVEN

Fast Science

7.1 Science in Supreme Emergencies

The tortoise taught us that tenacity triumphs, while the hare's hubris handed him heartbreak. Routine science is a tortoise. Is fast science a hare? A supreme emergency can motivate scientists to cut corners to quickly find a response. The epidemiological modeling at the start of the COVID-19 pandemic is a portentous example: This work was done rapidly, based on empirical assumptions that were soon criticized, promulgated without peer review, presented quickly to policy-makers, and formulated in terms of maximal urgency to elicit specific policies. The resulting response—lockdown—was one of the most impactful science-informed policies in decades. The putative warrant for violating the principles and practices of routine science was obvious: Many millions of people could otherwise die. The magnitude, imminence, and plausibility of a supreme emergency warrant fast science. Yet the hare was too confident in his talents, so he took a nap and lost the race. My aim in this chapter is to articulate and defend two principles to keep fast science more like the tortoise than the hare.

In chapter 1, I introduced the notion of justificatory norms. Routine science follows a range of principles and practices that render it relatively reliable. What exactly those justificatory norms are has been a matter of dispute: Popper held that genuinely scientific theories should be falsifiable; Merton called his four principles communism, universalism, disinterestedness, and organized skepticism; Longino emphasized the importance of criticism, including the need for recognized avenues of criticism, shared standards, equality of intellectual authority, and responsiveness to criticism. However one characterizes the justificatory norms of science in such general terms, many routine scientific practices are meant to instantiate those norms. These include long and rigorous traineeships, demanding standards for empirical methods and analysis, competitive allocation of research resources, critical peer review and modification of work in response to review, publication of results, and criticism of existing work. Routine science takes time. The organized skepticism of science entails that scientists usually view their results tentatively. Moreover, most routine science is intended for peers rather than the public or policy-makers. Of course, many cases of routine science are not ideal, but the satisfaction of deontic justificatory principles and practices nevertheless exemplifies routine science at its best. Fast science violates some of the justificatory norms of routine science.

The putative warrant to violate justificatory norms in fast science is compelling. If a threat is catastrophic, imminent, and plausible, we should quickly learn as much as possible about it and how to intervene. In response to great threats, doing fast science may be better than doing no science, even if the fast science violates some justificatory norms. Drawing on Walzer's (1973) notion of "dirty hands," which holds that political and military leaders can violate routine moral codes in response to a supreme emergency, Birch (2021) argues that scientific policy guidance can violate routine norms during periods he calls in extremis—times when many people are in immediate danger and there is no plan in place to manage the threat. Birch argues that scientists should usually limit themselves to "normatively light advice" where they offer no specific policy recommendation, but when in extremis, scientists can provide "normatively heavy advice" where scientists do offer specific policy recommendations. Though I am unconvinced by Birch's specific claim regarding normatively heavy advice during an emergency, a similar argument can warrant fast science. Engaging in and acting on fast science is warranted in the context of an extreme crisis. Walzer's claim that supreme emergencies can warrant the violation of moral principles might be contentious, but that such emergencies can warrant the violation of justificatory norms is much less so. White et al. (2022, 12) agree, claiming that normal epistemic standards can be violated during a supreme emergency.

I characterize a supreme emergency as being constituted by a *threat tripod* of magnitude, imminence, and plausibility. The magnitude gives some indication of the scale of the catastrophe if we were to not engage in or act on fast science. Imminence indicates the temporal restrictions within which fast science can operate. Walzer's (2000) notion of supreme emergency was based on a threat's magnitude and imminence, yet the threat tripod is incomplete without plausibility. This is because threats have varying degrees of plausibility, and, like the estimated magnitude of a threat, a threat's plausibility indicates what things would be like if we were to not engage in or act on fast science (if the plausibility of a threat is low, then things would probably be much the same as they are now without fast science, whereas if the plausibility of a threat is high, then things could be very different than they are now without fast science). The plausibility leg of the tripod can be understood as a probability of some minimal degree arising from uncertainty: In the spring of 2020, the emerging pandemic was a plausible catastrophe, but a plague of zombie unicorns was not. The magnitude and plausibility of a threat are standard considerations in risk assessment, and a supreme emergency has the added feature of imminence.

The three legs of the threat tripod must all be tall enough for a threat to be a supreme emergency. If a threat is trivial but imminent and plausible—say, tomorrow there will be no milk in my fridge for tea—then the magnitude leg of the tripod is too short and so the tripod does not stand, there is no supreme emergency, and no fast science is warranted. If a threat is far off in time yet catastrophic and plausible—say, in a few billion years, the sun will turn into a red giant and engulf Earth—then the imminence leg of the tripod is too short and so the tripod does not stand, there is no supreme emergency, and no fast science is warranted. If a threat is implausible yet catastrophic and imminent—say, tomorrow a gamma-ray burst from our solar system will strike Earth and kill all life—then the plausibility leg of the tripod is too short and so the tripod does not stand, there is no supreme emergency, and no fast science is warranted (low probability threats of a catastrophe should, of course, be studied by routine science). Yet if the three legs of the threat tripod are tall enough, the supreme emergency warrant says that catastrophe is likely coming soon, and so we should engage in and act on fast science.

The supreme emergency warrant asks us to forgive science for being more hare than tortoise and to put aside misgivings about the potential harms, both epistemic and practical, of violating the justificatory norms of routine science. But the supreme emergency warrant is incomplete because it says nothing about how hare-like fast science can be. Here lies a worry, because, as above, fast science departs from routine science by violating the deliverances of the Leibniz procedure, and, as we saw in chapters 1 and 2, satisfying those justificatory norms contributes to the reliability and trustworthiness of routine science—in fast science, scientists are invited to violate some of those deontic justificatory norms. But this cannot be an invitation for scientists to consult tea leaves, tarot cards, and mystics. There must be limits to fast science, especially in a context in which policy-makers claim to "follow the science." In this chapter, I articulate and defend two principles for guiding and assessing fast science.

The first of these is the Principle of Similarity, which says that fast science should be as similar as possible to routine science while satisfying the temporal and material constraints of the supreme emergency that motivate the fastness of the fast science in the first place. The supreme emergency warrant for fast science does not mean that anything goes but that scientists can cut some corners of routine science while satisfying the justificatory norms of routine science as much as possible. This principle is straightforward and, as I note in §7.3, so basic that it needs little argument (though in §7.4, I describe a recent case in which the principle was flouted to some degree). The second is the Principle of Independence, which says that violations of justificatory norms when developing mitigation strategies for a threat should not be the same as or similar to violations of the justificatory norms that were used to detect the threat (in §7.3, I articulate this principle in more detail and explain what "same" or "similar" means in this context). That is, scientific work that is intended to inform our choice of intervention should not depend on the same violations of justificatory norms that were used to estimate the magnitude, imminence, and plausibility of the threat itself.

Similarity and independence are, in general, graded notions. An instance of fast science satisfies the two principles to some degree. And, of course, the magnitude, imminence, and plausibility of a threat are graded properties—the taller the threat tripod, the greater the supreme emergency context of fast science. So a supreme emergency involves considering graded properties that speak in favor of the violation of justificatory norms (the threat tripod) and graded properties that speak against that violation (the two principles). This evaluation can occur both ex ante and ex post. Assessment ex ante could take the scientist's perspective and ask, Given the estimation of the threat tripod, to what degree should I violate justificatory norms? Assessment ex post could take the decision-maker's perspective and ask, Given the estimation of the threat tripod, to what extent should my choice of intervention be based on the results of scientific work that violated justificatory norms?

After further characterizing fast science in §7.2, I defend the two principles for fast science in §7.3. I describe two famous episodes of fast science in §7.4—the Manhattan Project and COVID-19 pandemic modeling—and I use the two principles to assess these episodes. Contexts of fast science are also contexts of decision under uncertainty, so, in §7.5, I describe how fast science relates to theories of decision under uncertainty and great threat, particularly versions of a precautionary principle. In §7.6, I conclude that fast science cannot be a tortoise, but it should not be too much like a hare.

7.2 Fast Science

Although there is an obvious temporal dimension to fast science, it is not merely the pace of scientific work that distinguishes fast science from routine science. Much routine scientific work is carried out in conditions of great temporal pressure resulting from priority races and the tenacity of ambitious scientists, as occurred with the discovery of the double-helical structure of DNA by Watson and Crick. As noted in chapter 6, the reward structure of routine science can motivate hasty work (Heesen 2018). Conversely, some instances of scientific work with many of the features of fast science occur over several years, as happened with the Manhattan Project (an example I discuss in §7.4). What distinguishes fast science from routine science is the violation of the justificatory norms of routine science, in response to a supreme emergency.

There is a set of justificatory principles and practices that makes routine science relatively reliable, and that is the basis for the objectivity of science, as argued in chapters 1 and 2. Moreover, particular justificatory practices are domain-specific: Testing the effectiveness of a new drug might require randomized trials, while predicting the future location of an astronomical object might require simulations; some scientific communities use practices such as peer review, while others share findings online. These domain-specific justificatory practices vary on even more fine-grained properties—for example, a randomized trial can involve one hundred subjects or one hundred thousand subjects.

To understand fast science, we do not need a complete list of all the justificatory norms of routine science, nor must we suppose that the distinction between fast science and routine science is sharp. It is enough to say that routine science is characterized by justificatory norms, and fast science is policy-guiding science that departs from some or many of those justificatory norms and receives putative warrant for that departure based on the threat tripod of a supreme emergency. Fast science is defined in terms of the deliverance of the Leibniz procedure described in chapter 1: It involves the violation of some relevant justificatory norms, and that violation is putatively warranted because of the great threat posed by a supreme emergency.

There is a recent movement dubbed "slow science," which is characterized by careful, methodical work, is motivated by curiosity rather than practicalities, and resists the publish-or-perish model of routine science (Stengers [2013] 2018). Though fast science and slow science naturally differ in many important respects, fast science is best understood by reference to routine science rather than slow science.

The fastness of fast science—and, to repeat, I do not strictly mean the temporal dimension—can occur at two distinct phases. The first I call the "detection" phase, which is the scientific work that provides estimates of the magnitude, imminence, and plausibility of a threat. An assessment of whether a scenario is a supreme emergency may depend on fast science in the detection phase. I call the second phase the "intervention" phase, which is the scientific and technical development of threat mitigation strategies. Regarding the episodes of fast science discussed in §7.4, in the Manhattan Project these two phases were to a significant extent temporally and scientifically distinct, while in the COVID-19 modeling during the spring of 2020, they were to some extent collapsed into the same scientific work.

Some usual scientific work is like fast science in that it violates some of the principles and practices of routine science, but without the supreme emergency context. Elliott and McKaughan (2014) describe the scientific assessments of restored wetlands, which must be "authoritative, cheap, and quick," are not peer-reviewed or formally published, and are policy-guiding (see also Steel 2016a). While the practical context of assessing restored wetlands may provide a putative warrant for violating justificatory norms, that practical context is clearly not a supreme emergency. Appeals to science in legal contexts are often similar (see, for example, Miller 2016). In this chapter, I am concerned only with fast science in response to catastrophic, imminent, and plausible threats.

Fast science can be understood by considering inductive risk. The inductive risk argument states that a scientist can make a mistake in two basic ways: She can accept a hypothesis when it is false or reject a hypothesis when it is true (for more detail see chapter 3). Both kinds of errors have practical consequences, and, for a particular hypothesis, assessing these consequences influences our evidential thresholds for accepting or rejecting the hypothesis and acting accordingly. In some scenarios, we might be willing to accept and act on a hypothesis with very little support because the consequences of not acting could be catastrophic. Fast science has three features that make inductive risk considerations particularly important. First, the uncertainties of fast science are typically greater than the uncertainties of routine science. Although routine science has inductive risk, this is in the context of scientific work that can satisfy the justificatory norms of routine science to a greater degree, while fast science by definition involves violating some of the justificatory norms and hence an increase in the uncertainty of its results. Second, one of the usual responses to the inductive risk argument in routine science is to conclude that values influence scientific reasoning, and so scientists must carefully manage that value influence by, for example, being transparent about the influence of values (Elliott 2022), being selective about which values are involved (Kourany 2010), appealing to democratic values (Schroeder 2021), developing shared conventions (Wilholt 2013), and enriching the diversity of value-perspectives (Longino 2004). That is all useful, but it takes time. Routine science has that time, whereas the fastness of fast science can hinder these value-management strategies. Third, the threats in fast science are catastrophic in magnitude, while the risks in routine science can include harms of any scale, which may be catastrophic (say, if our climate models severely underestimate the impact of carbon emissions), yet many such risks are, in comparison to catastrophe, minor in magnitude.

Friedman and Šešelja (2023) argue that higher-order evidence can be important in assessing scientific controversies, particularly in the context of fast science. By fast science, the authors mean "application-driven research confronted with an urgent need to accept or reject a certain hypothesis for the purposes of policy guidance, aimed at addressing a significant pending social harm" (938). Important for their definition, delaying the assessment of a hypothesis of concern to allow time to gather more evidence may be harmful, and hence the context of fast science requires endorsing (or rejecting) a hypothesis without waiting for further evidence. This aligns with my view of fast science. As I argued in chapter 3, in routine science, it is typically possible to collect more and better evidence for a hypothesis, and not doing so may amount to violating a justificatory norm. The question I address concerns the putative warrant of fast science, while Friedman and Šešelja's question is about the importance of higher-order evidence for thinking about how to respond to scientific disputes during fast science.

Shaw (2022a, 108) defines "urgent science" as follows: "A research proposal is urgent if there is a practical or moral reason to need a result within a specified timeline and the research can realistically be carried out within that timeframe" (see also Shaw 2022b). Shaw discusses this idea in terms of the pursuit-worthiness of scientific research programs. The notion of urgent science is clearly related to my concept of fast science, as fast science gets its putative warrant based on a practical or moral reason to need quick results. Yet Shaw's notion of urgent science is broader than my concept of fast science, as I am specifically concerned with scenarios of supreme emergencies.

Like a spy in the dark alleys of an old European town, threats can sneak up on you. Information about the magnitude, imminence, and plausibility of a threat can trickle in through various channels that might resemble routine science—until you wake one morning to find yourself in a supreme emergency. Developments in atomic physics in the 1930s culminated in physicists recognizing that atomic energy could be harnessed to build massive bombs and that the Germans were probably pursuing this. Similarly, in the first three months of 2020, a few reports emerged about the spread and lethality of the new coronavirus, which culminated in the dire predictions of epidemiological modelers in March 2020 (White et al. 2022). Because some threats can gradually emerge, the detection phase of fast science might require violating some routine scientific practices. The fastness of the intervention phase is, in turn, putatively warranted by the supreme emergency and by the fact that, based on the novelty of the threat, there is a dearth of results from existing routine science that can be relied on for developing interventions.

One might ask if there is a potential circularity regarding the warrant of the detection phase. The task of the detection phase is to assess the threat tripod, but establishing the threat tripod is precisely what warrants the violation of justificatory norms. Yet both the detection phase and the intervention phase of fast science are literally phases and not moments. I noted previously that supreme emergencies can sneak up on us; both the lead-up to the Manhattan Project and the early months of the COVID-19 pandemic exemplified this. The earliest signals of a threat can be (and perhaps often are) vague and may indicate danger even if all three legs of the threat tripod are not standing tall, but this can still require further rapid work in the detection phase. Moreover, violations of justificatory norms in the intervention phase plausibly require greater warrant with respect to the threat tripod, relative to the detection phase, since the intervention phase has implications for policy while the detection phase does not.

FAST SCIENCE   191

To summarize, fast science:

i. Violates justificatory norms of routine science
ii. Guides policy responses to a supreme emergency
iii. Has two phases: detection and intervention

The supreme emergency warrant adds: (i) is putatively warranted by (ii). Some instances of routine science, unfortunately, also have feature (i) — indeed, the satisfaction or violation of justificatory norms is the basis of scientific evaluation I discussed in chapter 2. Thus, one may be unable to easily demarcate instances of fast science from instances of sloppy routine science based only on the violation of justificatory norms. If one were motivated by this aim of demarcation, they could simply check whether feature (ii) is present along with feature (i). In the following section, I ask about the putative supreme emergency warrant of fast science. Since the heart of science is justification, the violation of deontic justificatory principles and practices should be carried out cautiously and assessed critically. This chapter sets out two principles to consider when assessing fast science.

7.3 Similarity and Independence

Consider this case:

> *Asteroid Impact*
> Sasha consults astronomical observations and established theory and predicts that an asteroid will soon strike the earth, causing incredible destruction. In a panic, she does some quick back-of-the-envelope calculations to ascertain the most effective mitigation strategy. She writes a report for policy-makers and holds a press conference urging immediate implementation of her recommended intervention.

In the intervention phase of *Asteroid Impact*, Sasha's method departs from routine science. Sasha violates the Principle of Similarity.

The Principle of Similarity is intuitive and so straightforward that it hardly needs an argument. It is worth stating, however, because, as we will see in §7.4, a prominent example of fast science during the COVID-19 pandemic did not adequately satisfy the principle. If a threat's magnitude, imminence, and plausibility are sufficiently severe, we might be willing to violate justificatory norms, but there are limits: The scientific work in fast science, both in the detection phase and particularly in the intervention phase, should not stray so radically from routine science that the resulting work becomes highly unreliable. If such work excessively violates justificatory norms, it should not be relied on for policy, as in *Asteroid Impact*. Yet similarity is a graded and multidimensional notion. Any reason that the deliberators in the Leibniz procedure might consider for admitting a justificatory norm into routine science is also a defeasible reason for an instance of fast science to satisfy that justificatory norm. The very idea of warranted fast science supposes that supreme emergencies provide at least some warrant for violating at least some of the justificatory norms of routine science, yet the standing reasons for those deontic norms remain and should constrain the resulting fast science methods as much as possible, within the temporal and practical bounds of the threat tripod.

Consider a broad principle like Longino's (1990) norm that science should be responsive to criticism. An episode of fast science could involve a group of scientists with no critical feedback internal to the group and no engagement with external critics, before they present their work to policymakers. Such fast science would be very dissimilar to routine science on this broad norm of responsiveness to criticism. Another group doing similar work could give each other critical feedback and seek critical input from outside experts. Such fast science would be more like routine science on this norm and would thus better satisfy the Principle of Similarity.

The Principle of Similarity can also be assessed by reference to fine-grained justificatory practices. Suppose that to properly test a new drug in a randomized trial in a routine science context, 100 subjects are necessary. Suppose also that a threat tripod entails that there is not enough time to test 100 subjects: A trial with 99 subjects would satisfy the Principle of Similarity more than would a trial with 98 subjects, and a trial with 98 subjects would satisfy the Principle of Similarity more than would a trial with 97 subjects, and so on.

Walzer (1973) famously argued that supreme emergencies can warrant the "dirty hands" of political and military leaders who act immorally to save their community by, for example, bombing civilians or torturing informants. Birch (2021) applied Walzer's argument to conclude that scientists can violate scientific norms in supreme emergencies. Walzer's argument is controversial; torture, for example, is unreliable, and there is a potential contradiction to the idea that immoral behavior can be morally justified. In any case, there is an important difference between the supreme emergency warrant of engaging in and acting on fast science and the supreme emergency warrant of dirty hands. The dirty hands argument is supposed to warrant behavior that is drastically *unlike* the sort of behavior that is permitted by routine moral principles, and more drastic departures from moral behavior may be more effective in achieving one's aim. Yet in supreme emergency scenarios, the more that fast science violates the justificatory norms of routine science, the less reliable it is, so the resulting recommended intervention is less likely to be effective. When the legs of the threat tripod are sufficiently tall, immoral behavior may be warranted (though that is controversial), but unscientific behavior is not warranted.

Let us now consider the second principle for assessing fast science. The Principle of Independence says that the scientific work that guides our choice of intervention should not involve the same violations of justificatory norms of the scientific work that was used to assess the threat tripod. I offer two distinct arguments for the Principle of Independence. The first is based on considerations of reliability, like the argument for the Principle of Similarity. The second is based on a second-order principle that holds that violations of norms cannot be self-justifying.

The Principle of Independence is meant to guarantee that whatever specific form of unreliability exists in the detection phase of fast science is not inherited by the intervention phase. Recall that what defines fast science is the violation of justificatory norms of routine science. The potential unreliability of a method used in fast science cannot serve as a justification for disregarding its results or failing to act on those results because the reliability of all fast science should be questioned—because it is fast science, some of the justificatory norms have been violated. Yet we have granted the supreme emergency warrant of engaging in and acting on at least some fast science. There may be an estimation of a severe threat tripod in the detection phase, but even if the Principle of Similarity is satisfied, the fast science in this phase could be unreliable in a particular way, and that particular form of unreliability could arise in the intervention phase. Errors in estimating the threat tripod could be compounded by the same errors in assessing mitigation strategies. The Principle of Independence is meant to guarantee that this does not occur.

The Principle of Independence holds that, so long as it does not come at the cost of decreased reliability, the scientific work that guides our choice of intervention should not depend on the same violations of justificatory norms used to assess the threat tripod. But "same" here is ambiguous: It could mean the same token violation or the same type of violation. In §7.4, I give an example where the same token violations of justificatory norms occurred in both the detection and intervention phases. In the following arguments, I focus on token-sameness, as it is particularly problematic, yet these arguments also apply, though perhaps less forcefully, to type-sameness.

To illustrate the importance of the Principle of Independence, consider another case:

Pandemic

Lilia simulates the spread of a new infectious disease using her in-house software, which has been developed over years, is informed by past epidemics, and has been the basis of many peer-reviewed publications. She uses the scant data she can access to set the values of the parameters in the simulation. Her results suggest that the disease will cause a terrible catastrophe. Using the same model, she also simulates mitigation strategies, and her results suggest that her favored strategy will save millions of lives. She writes a report for policy-makers informing them of her findings and holds a press conference urging immediate implementation of her recommended intervention.

The fast science in *Pandemic* satisfies, as much as could be expected given the supreme emergency context, the Principle of Similarity in both the detection and intervention phases. In that regard, this case is superior to *Asteroid Impact*. But, unlike *Asteroid Impact*, *Pandemic* involves the very same token methods, assumptions, and potential sources of unreliability in both the detection and intervention phases, so the Principle of Independence is not satisfied. Now suppose that the scant data Lilia uses to set her parameters are unreliable, and those parameter values cause the simulation to grossly overestimate the speed of disease spread and the lethality of the disease. That overestimation, in turn, contributes to a significant overestimation of how many lives would be saved if Lilia's favored mitigation strategy were implemented. This is not a departure from the Principle of Similarity because the work is as close as possible to what could have been achieved in routine science given the temporal constraint. The problem is that the specific source of unreliability in the intervention phase — namely, the value of the simulation parameters — is the same token source of unreliability in the detection phase. If the Principle of Independence were satisfied, that would have been avoided.

This rationale is similar to diversity of evidence arguments discussed in philosophy of science, which hold roughly that the greater the variety of a set of evidence supporting some hypothesis, the more likely it is that the hypothesis is true (see Kuorikoski and Marchionni 2016; Schupbach 2018; and Claveau and Grenier 2019). In these arguments, the multiple lines of evidence are about the same hypothesis or phenomenon, while the two phases of fast science are about two distinct phenomena: the estimation of the threat tripod in the detection phase and optimal mitigation strategies in the intervention phase. There is a risk of error in both the detection and intervention phases, and just as it is good to spread risk across multiple investments in a portfolio, it is good to spread risk across the two phases of fast science. The investment portfolio analogy helps show that token-sameness error risk is worse than type-sameness: A very risky investment strategy is to put all of one's resources into, say, shares of one particular technology company, while a less risky strategy, though risky nonetheless, is to invest in multiple token technology companies; yet the latter strategy involves investing in a single type of company, so an even less riskier strategy is to invest in multiple types of companies from different industries.

When the Principle of Independence is satisfied, the results of the intervention phase can serve as a robustness check on the results of the detection phase. Since we have granted that some reliability-enhancing features of routine science can be violated in both phases, such a robustness check can be valuable in minimizing error. In *Pandemic*, for example, suppose that during the intervention phase, researchers did not use only the model to evaluate interventions but also performed a cluster randomized trial in which some communities were randomly allocated to particular mitigation interventions and others to no intervention. The results of the intervention phase, which satisfies Independence, would serve as an independent test of the results of the detection phase. That is because the control group of such a study would provide evidence relevant to the predictions that the model made about a scenario in which policy-makers do nothing, and the intervention groups would similarly provide evidence relevant to the predictions made by the model in the various intervention scenarios. If the model in the detection phase predicted that roughly x number of people will get infected with no intervention but roughly y number of people will get infected if we close schools, and if an extrapolation from the trial roughly confirms those values of x and y in the two respective scenarios, then policy-makers would have much better epistemic justification regarding school closure. But if the Principle of Independence is not satisfied, such a robustness check is not possible because both the detection and the intervention phases share the same risks of error. (One might worry that performing a randomized trial would take too much time in the context of a supreme emergency; the point here is simply to articulate the epistemic benefits of satisfying the Principle of Independence. In any case, during the COVID-19 pandemic, scientists did perform trials like the one suggested here, such as the Bangladesh mask study discussed in chapter 4.)

Here is a slightly different way to make this point. The fastness of fast science is putatively warranted by a supreme emergency. Estimation of the threat tripod occurs in the detection phase. The results of the detection phase, in turn, offer putative warrant to the fastness of the intervention phase. Yet the detection phase has a risk of error, and so the warrant of the fastness of the intervention phase is threatened. If the intervention phase does not share the same risks of error as the detection phase, then the results of the intervention phase can provide independent support for the results of the detection phase. Because it is the results of the detection phase that warrants the need for the intervention phase, the result is a mutual warrant between phases, when Independence is satisfied. If Independence is violated, however, this mutual warrant is not available.

Let us turn to the second argument for the Principle of Independence. In chapter 1, I argued that the justificatory principles and practices of routine science can be formulated as justificatory norms. I described the normative basis of justificatory norms (and I argued that there is more to the normative basis of justificatory norms than merely their truth-conduciveness or reliability, but I simplify that point for the discussion here). Fast science violates justificatory norms of routine science. Since they are pro tanto norms, their violation may be justified, perhaps on grounds of feasibility, cost, ethical constraints, or a supreme emergency. Yet the violation of any norm, pro tanto or not, scientific or not, fast science or not, cannot be warranted by the very instance of norm violation needing warrant. Norm violation cannot be self-justifying.

The viciousness of self-justification of norm violation may be more clear in nonscientific contexts. Imagine Lilia is breaking the speed limit as the only driver on a road passing a school. A police officer pulls her over and tells her she was breaking the law by speeding, and she says that she was just driving at the average speed of all cars on the road at the time. Imagine Neil is caught meeting his lover during a lockdown in which nonhousehold social contacts are prohibited because they could cause harm. He says that his meeting was fine because it brought him a lot of pleasure. Imagine a devoted Kantian who sincerely believes that lying is never permissible and then lies about the permissibility of lying by appealing to its utility-maximizing consequences only to justify to himself that very lie. Lilia's speeding, Neil's tryst, and even the Kantian's lying may have been warranted violations of norms, yet those warrants must be based on something other than the very act in question. Perhaps Lilia was bringing her ill child to a hospital, so she needed to speed—that, however, would not be a self-justification of norm violation, as it would appeal to something other than the act itself or features of the act as grounds for violating the norm (see Gert [1998], chapter 9, for a discussion of norm violation).

Here is a general statement of what I am concerned with:

Self-justification of norm violation
S does action A, which violates norm N, and S cites features of A as warrant for violating N

Features of A that one might cite (absurdly) in an attempt to self-justify a norm violation could include its statistical normality (Lilia's speeding), the benefits that it brought to S (Neil's tryst), or its sheer permissibility (the Kantian's lying). In each case, the supposed self-justification is laughable. To be sure, citing features of A can be relevant to assessing N's status *as a norm*. Suppose A resulted in no harm, and a compelling case could be made that actions like A rarely result in harm, but they bring many benefits. This could challenge the status of N as a norm, though it would not justify S's behavior in that instance. Moreover, this consideration is not applicable to fast science since fast science involves violating a justificatory norm and not challenging whether that norm is indeed justificatory.

Let us consider a scientific example of self-justification of norm violation: A scientist who does A, which violates N, is challenged about her violation of N, so she cites features of A, particularly the outcome of doing A. For example, Sasha is testing a new fertilizer for sunflower plants, and her statistician tells her that to get sufficient statistical power, she will need to randomly allocate one thousand sunflower plants each to a fertilizer group and a control group. But after getting data on only twenty sunflower plants, Sasha stops the trial, analyzes the data, and gets results R, which suggest that the sunflower plants grew taller than she expected with the fertilizer. The statistician criticizes Sasha, stating that because her trial did not include enough sunflower plants, R is unreliable. Sasha responds by citing the large effect size of R as grounds for violating N. Yet the statistician considers R to be unreliable precisely because it violated N and thus considers R no grounds at all for violating N.

The self-justification of norm violation in the cases above is synchronic: S cites features of A as a warrant for the violation of N by that very act A. But fast science is defined as occurring in two stages, and if those stages are temporally distinct, then the self-justification of norm violation would be diachronic: S cites features of A_1 at time T_1 as a warrant for the violation of N via A_2 at T_2. The cases above can be given a diachronic structure, and the putative self-justification of such cases remains laughable on a diachronic structure. Neil could have met his lover on several occasions, and on the occasion when he is finally caught, he could offer as a putative warrant the pleasure that his prior trysts brought him. That would be as absurd as the synchronic cases above. (Regardless, as we will see below, in the real case that I use to illustrate the violation of Independence, the putative self-justification of norm violation is synchronic.)

Identifying which justificatory norms have been violated in fast science may not always be obvious, and it may not always be easy to tell if the Principle of Independence is satisfied. Some violations of justificatory principles and practice of routine science are transparently discernible, while others are harder to discern.

To sum: I consider fast science to be composed of two phases: estimation of the threat tripod and assessment of interventions for the threat. The Principle of Independence states that violations of the justificatory norms of routine science in the intervention phase should not depend on the same violations of justificatory norms in the detection phase, so long as this does not come at the cost of decreased reliability. In §7.1, I argued for the supreme emergency warrant for violating justificatory norms, including in the detection phase. However, violations of justificatory norms in the intervention phase should not be given the supreme emergency warrant if the assessment of the threat tripod was based on the same violation of justificatory norms in the detection phase, as that would amount to self-justification of norm violation.

To be clear, both principles are important only as a way to minimize error and maximize the chance of good outcomes from whatever policy is deployed. They are defeasible, particularly given a significant threat tripod. Recall that in §7.1, I argued that the magnitude, imminence, and plausibility of a threat are all graded properties of a supreme emergency, and the greater they are, the more the two principles defended here can be violated. Moreover, there may be an interaction between Similarity and Independence: If Similarity is satisfied to a very large degree in the detection phase, it may be acceptable to violate Independence in the intervention phase (because the detection phase methods were fairly reliable). It might even be more reliable to violate Independence compared with using a very unreliable method that satisfies Independence; however, if Similarity is grossly violated in the detection phase, then it is very important to satisfy Independence in the intervention phase (precisely so that the unreliability in the detection phase does not carry over to the intervention phase).

There may be other principles relevant to fast science, particularly of a nonepistemic nature, depending on the details of a supreme emergency. Suppose a supreme emergency involves a threat to human reproductive capacities, and the birth rate plummets to near zero. Scientists might try to maximize the probability of discovering an effective intervention and, depending on the type of threat, they might choose to care less about false-positive discoveries and care much more about avoiding false negatives on broadly inductive risk considerations. By doing so, they would maximize the chance of a good outcome despite increasing the frequency of a particular kind of error (false positives).

7.4 The Tortoise and the Hare

In this section, I detail two episodes of fast science referred to earlier. The first is more tortoise than hare, the second is more hare than tortoise.

Manhattan Project

An example of fast science that arguably satisfies both principles is the Manhattan Project. Although it has been described as "big science" (Hughes 2003), the Manhattan Project also had many features of fast science: The work was done with extreme secrecy, so there was minimal peer review in scholarly publications, and it was performed under intense time pressure with a clear action-oriented goal. The magnitude, plausibility, and imminence of the threat—Germany being the first to develop atomic bombs—became clear to émigré physicists by the late 1930s, leading to the 1939 Einstein–Szilard letter warning President Franklin D. Roosevelt about the possibility of "extremely powerful bombs of a new type." This was the detection phase.

During the Manhattan Project, many scientists were opposed to the secrecy and compartmentalization of their work. For example, in 1944, Leo Szilard wrote to Vannevar Bush, one of the initiators of the Manhattan Project, that "decisions are often clearly recognized as mistakes at the time when they are made by those who are competent to judge, but ... there is no mechanism by which their collective views would find expression or become a matter of record" (cited in Rhodes 1986, 508). When Edward Teller tried to convince fellow physicist Niels Bohr that nuclear research should be kept secret during the war, Bohr responded by insisting that secrecy should not be a part of science (Rhodes 1986, 294). Contrary to routine science, the atomic physics of the Manhattan Project involved no peer review of the usual sort, no publication, and extreme secrecy, all in the service of a military technology.

The intervention phase was, of course, directed toward building atomic bombs. The Manhattan Project involved an intelligence component that was tasked with gathering information about German atomic science and technology, but that work was substantively independent from the scientific and technical developments of the Manhattan Project. Obviously, results from the espionage could have contributed to the scientific and technical developments of the Manhattan Project, and these developments could have influenced the estimation of the magnitude, imminence, and plausibility of the threat. For example, failure to enrich uranium despite sustained effort would have suggested that the Germans were also having difficulty enriching uranium, and that in turn would influence the threat tripod, and success at enriching uranium would have had the opposite consequence. That additional evidence about the threat tripod would be especially compelling since the work during the intervention phase was quite different from that in the detection phase. Any violations of justificatory norms that occurred during the intervention phase were different from those that occurred during the detection phase. The Principle of Independence was satisfied.

The scientific work of the Manhattan Project was informed by decades of nuclear science and involved many of the most eminent physicists of the day. The director, Robert Oppenheimer, organized a conference focused on nuclear weapons in 1942. Huge sums of money were allocated to empirical, theoretical, and technical work, which was distributed over many research sites and subject to intense critical scrutiny, particularly among different teams. Although the Manhattan Project deviated from routine science practices, it included justificatory practices that served similar functions as the justificatory norms of routine science, and so the Principle of Similarity was to a large degree satisfied.

COVID-19 Models

In chapter 4 I described the Imperial College London COVID-19 modeling and resulting Report 9, which made dire predictions about the impact of the COVID-19 pandemic under a variety of policy scenarios ranging from taking no action to enacting various mitigation strategies (Ferguson et al. 2020). Birch (2021, 10) notes that Report 9 strongly recommended the policy of sustained lockdown as "the only viable option" (Report 9 describes lockdown as "the preferred policy option" and "the only viable strategy"). As noted in §7.1 and chapter 4, this work violated many of the justificatory norms of routine science. Critics of that work and the resulting policies argued that key parameters of the model were not well supported and that further sensitivity analyses would have demonstrated that the dire predictions were not robust to changes to those parameters (see, for example, Winsberg et al. [2020], a response from van Basshuysen and White [2021], and a subsequent rejoinder from Winsberg et al. [2021]). Two of the most important parameters were the reproduction number and the infection fatality rate, and while available empirical evidence suggested that the Imperial College London values for these parameters were probably inaccurate, subsequent work confirmed this (Northcott 2022). Report 9 was not peer-reviewed and was presented directly to policy-makers. Not long after the report's publication, its projections appeared to be grossly inaccurate (Winsberg et al. 2021). In short, the Principle of Similarity was, to a very significant degree, violated.

The scientists could have included a richer dataset to inform their parameters and could have performed more sensitivity analyses. One plausible reason they did not do so is that they were under extreme time constraints due to the nature of the threat, and, goes this thought, they simply could not wait for better data to inform their parameters or to perform more simulations. A delay in reporting the findings to policy-makers could have cost many lives. Yet specific predictions about the threat tripod and how to respond to the threat were the main purpose of Report 9. Consider, for example, the data in table 4 of the report, which suggests that in the United Kingdom the number of deaths over two years would be about half a million greater if no mitigation policies were deployed. Winsberg (2022) suggests that this modeling work was "self-recommending" because if scientists were required to gather more evidence to inform the model, it would delay policy action, leading to more deaths, as predicted by the model—thus there was no need to gather more evidence. This self-recommending aspect of the work is a violation of the Principle of Independence.

When the Imperial College London epidemiologists presented Report 9 to policy-makers—mid-March 2020—there were compelling reasons to infer that the threat of the pandemic was large, imminent, and plausible, particularly based on experiences in China and Italy. Yet the report's conclusions were profoundly dire and were presented with extreme precision and certainty in an authoritative context—the lead scientist, Neil Ferguson, had been a member of the Scientific Advisory Group for Emergencies from its first meeting about the pandemic in January 2020. Report 9 contributed specific and authoritative estimates of the magnitude, imminence, and plausibility of the threat of the COVID-19 pandemic. Using the very same model, with all its shortcomings, Report 9 provided very specific guidance on mitigation strategies. Figure 2 in Report 9 depicts the simultaneous assessment of the threat tripod and prediction of the effectiveness of various mitigation strategies, insofar as it includes simulation results for both "do nothing" and mitigation scenarios. This is a violation of the Principle of Independence.

This might be a common feature of epidemiological models, as van Basshuysen et al. (2021, 119) note, "Epidemiological models serve dual purposes: apart from their epistemic purpose of forecasting the course of an epidemic, they also serve the practical purpose of informing and guiding policymaking. These epistemic and practical purposes go hand in hand: on the basis of forecasts, policymakers can choose policies that are likely to prevent unwanted outcomes." That seems right to me. Yet if an epidemiological model has a significant flaw, then both of its dual purposes could be hindered by that flaw, the resulting policy could be significantly misguided, and the concern about self-justification of norm violation stands. Fast science, including epidemiological modeling, should strive to satisfy the Principle of Independence.

* * *

These two cases are meant as illustrations, and it would be helpful for future work to explore a richer range of cases. The development of the Russian Sputnik V COVID-19 vaccine, for example, might exemplify a case where Similarity was violated, because the vaccine was rolled out in Russia after only a phase 2 trial rather than the usual phase 3 trial (moreover, Sheldrick et al. [2022] reanalyzed the data from the trial and concluded that the data showed signs of manipulation). This work, however, plausibly satisfied Independence (since the violation of the justificatory practices in the detection phase had nothing to do with trials). It would be interesting for future research to explore a wider range of cases of fast science.

7.5 Caution and Intervention

Suppose there is a supreme emergency, and the resulting fast science satisfies both the Principle of Independence and the Principle of Similarity. Does that entail that whatever mitigation strategy is recommended in the intervention phase is best? It does not. Conversely, suppose the Principle of Independence and the Principle of Similarity are not satisfied in the intervention phase. Does that entail that whatever mitigation strategy is recommended in the intervention phase is not best? It does not. Fast science is still just science—it is not an inference rule or decision procedure.

For decisions in contexts of uncertainty, there are competing views about how agents should choose among various options. Some believe that such contexts require a standard cost-benefit analysis, even if estimates of the probabilities of outcomes are unavailable. Broome (2012, 29), for example, advises us to "stick with expected value theory," and Posner (2004) recommends standard cost-benefit analysis even in the face of a catastrophic threat. Some hold that the best approach is to postpone a decision until more information is available (see, for example, the discussion in Bradley and Steele 2015). Some advise deciding cautiously, with a focus on avoiding the worst possible outcomes. In this section, I ask how fast science relates to theories of decision under uncertainty.

Fast science can obviously give us information about a threat and inform policy-makers about options for intervention and their chances of success. Birch (2021), however, claims that fast science can do more. He states that in extremis contexts can warrant an "imperative to elicit a specific policy response" (90) because if that response were not elicited, the scientific advisor believes catastrophe would result. However, results of fast science grossly underdetermine the best intervention, for several reasons. First, the mitigation strategy recommended in the intervention phase of fast science could be extremely costly, or unfeasible, or have profoundly terrible unintended but foreseeable consequences, or have a low probability of success; the fastness of fast science entails that we should have some uncertainty about the probability of intervention success. Second, as Birch rightly notes, science advisors are not accountable to an electorate as democratically elected politicians are. Finally, science advisors have interests and values that may not reflect those of the broader public. To use Pielke's (2007) terms, scientists should be "honest brokers" rather than "issue advocates," and a supreme emergency would have to be truly catastrophic and imminent to override the above considerations (on Pielke's honest broker ideal, see Havstad and Brown [2017b]; Brown and Havstad [2016]).

Perhaps a more robust connection between fast science and policy response can be grounded on a precautionary principle. At the most general level, precautionary principles appeal to estimates of the magnitude of a threat and estimates of the costs of intervening on that threat and recommend avoiding worst-case scenarios. Various versions of the precautionary principle have been defended.

Using Rawls's maximin principle, Gardiner (2006) argues that precautionary interventions are warranted in response to a potential catastrophe when four conditions are met: (i) There is no information about the probabilities of the possible outcomes, or what little information there is ought to be sharply discounted, (ii) the intervention guarantees that the catastrophic outcome would not occur, (iii) the cost of the intervention is minimal, and (iv) alternatives to the intervention, including doing nothing, may result in the catastrophic outcome (see Steel [2015] for discussion). In a supreme emergency, (i) would arguably be met because the emergency constrains the amount of time science could devote to getting sufficient evidence to inform the probabilities. I believe that (iii) would rarely be met in responding to supreme emergencies, but regardless, (iii) is more applicable to regulatory contexts and can plausibly be discarded in a supreme emergency (Sunstein [2005, 382] argues that it should be discarded in any context). Condition (iv) could be met, though assessing the relevant counterfactuals would be epistemically demanding. However, I believe (ii) could rarely be met in supreme emergencies, and even if both the Principle of Similarity and the Principle of Independence were maximally satisfied, the residual uncertainty resulting from the fastness of fast science entails that scientists could not guarantee the success of their intervention (which is perhaps a general problem for Gardiner's approach since guaranteeing the success of an intervention is very demanding). Thus, on

Gardiner's version of the precautionary principle, many or perhaps all instances of fast science would not warrant precautionary intervention.

The sophisticated version of the precautionary principle defended by Steel (2015) suggests a different relation between precautionary reasoning and fast science. Like Trouwborst's (2006) version of the precautionary principle and others before it, Steel's is based on a tripod metaphor that differs from the threat tripod metaphor introduced in §7.1. Steel's precautionary tripod is based on a knowledge condition, a harm condition, and a recommended intervention. The knowledge condition pertains to how much information we have about a threat, and the harm condition is a function of the magnitude of the threat itself. These in turn recommend a specific intervention that should be aggressive to a degree proportional to the other two legs of the tripod. Like a camera's tripod, the three legs are adjustable, and its balance requires that the adjustment of one leg depends on adjustments of the others. This allows for useful flexibility in applying the precautionary principle in different kinds of scenarios.

Features of fast science influence the balance of Steel's precautionary tripod. The more the Principle of Similarity is violated in the detection phase, the less certain is the estimation of the plausibility, magnitude, and imminence of the threat. Thus the knowledge leg of Steel's tripod would be minimized, so by Steel's proportionality requirement, the recommended intervention should be less aggressive than if the Principle of Similarity were better satisfied. The more the Principle of Similarity is violated in the intervention phase, the more uncertainty there is about the intervention's effects. If that uncertainty is great enough, then the very same precautionary principle that favors the intervention could also speak against the intervention (and that would in turn violate another requirement of Steel's precautionary principle, namely, the consistency requirement, which holds that an intervention should not be recommended against by the same precautionary principle that recommended in favor of it).

Even if a decision procedure—precautionary or not—justifies implementing a particular intervention, that justification should have a temporal restriction because, during the time the intervention is being implemented, better-quality scientific work can be performed, the results of which could then recommend against the intervention (a point argued by Winsberg et al. [2020] and White et al. [2022] in the context of COVID-19 lockdowns).

Some threats are great enough to eliminate all life on Earth, or at least all human life. Bostrom (2013, 15) defines an "existential risk" as a risk that "threatens the premature extinction of Earth-originating intelligent life or the permanent and drastic destruction of its potential for desirable future development." Examples include those mentioned above—asteroid impacts, nuclear war, pandemics—as well as artificial intelligence, climate change, hostile aliens, and gamma-ray bursts, among many others. If these threats could provide a supreme emergency warrant for engaging in and acting on fast science, scientists could violate the justificatory norms of routine science on a large scale, given the number of existential threats and their magnitude. This would have harmful epistemic consequences, as our study of these threats and corresponding intervention strategies would become less reliable and could have harmful practical consequences, as interventions could be costly and failed interventions could be catastrophic. Existential risks clearly involve threats of a great magnitude, and they generate interest because they are plausible (some more so than others). Yet they are probably less imminent than, say, the COVID-19 pandemic was in the spring of 2020. Hundreds of millions of years will likely pass before a large enough and close enough gamma-ray burst ends life on Earth. One might say that imminence should be construed both in terms of how far off in time a threat is as well as how much time is required to develop an effective intervention against that threat; if so, the above existential risks may very well be imminent. But estimating the time needed to develop an effective intervention against threats like gamma-ray bursts or hostile aliens would be purely speculative, in part because it is impossible to predict the future surprises of science. Putting this nuance aside, typical existential risks do not warrant fast science. The existential risk of climate change is much more imminent, but there is still time—decades, at least—to perform plenty of routine science before a climate apocalypse ends it all.

7.6 Conclusion

We have just emerged from a time when an episode of fast science turned our lives upside down. With a few exceptions, philosophers of science at the time had little to say about the COVID-19 pandemic and policy response. Our discipline, slow and steady, is accustomed to studying routine science, a tortoise watching tortoises. Yet science is sometimes more like a hare.

The heart of science is justification—specifically, the deontic justificatory norms delivered by the Leibniz procedure. We must do all we can to keep the heart of science healthy. Thus, violations of justificatory norms in science should be carefully scrutinized.

I have argued that supreme emergencies can warrant engaging in and acting on fast science. If a threat is catastrophic, imminent, and plausible, then scientists can cut corners to quickly recommend interventions. Yet fast science should be as close as possible to routine science in its satisfaction of justificatory norms, and those justificatory norms that are violated when developing an intervention for a supreme emergency should not be the same as those that are violated when estimating the threat tripod. Fast science cannot be a tortoise, but it should not be too much like a hare.

CHAPTER EIGHT

Timeless Truths

8.1 Introduction

One of the most dangerous beliefs about science is that all scientific findings are tentative and very likely to be overturned in the future. I call this the *provisionality thesis*. If only a few cranks and conspiracy theorists believed the provisionality thesis, it would not be so troubling. But a wide range of scholars who have devoted their careers to the historical, sociological, and philosophical study of science assert the provisionality thesis, as do many scientists. This is surprising, since the provisionality thesis is false.

The provisionality thesis is dangerous because it supports a general distrust of science. If all scientific findings are provisional, then we have less reason to rely on science and support science. Consider an analogy: If I told you that the reliability of this model of airplane is provisional, you would probably be less inclined to fly in that model. There are, of course, good reasons to withhold trust about particular areas of science. Research on pharmaceuticals, for example, is not very trust-inspiring because much scientific evidence about pharmaceuticals is withheld from the public. Psychology has recently been undergoing a "replication crisis" in which many prominent findings cannot be replicated, leading critics to conclude that those findings were spurious. I write this book as the world recovers from the COVID-19 pandemic, during which grossly unreliable science was relied on to inform influential policies. Nevertheless, science has been and will continue to be the best way to learn about the complex world around us. Despite that, distrust in science appears to be widespread, and such distrust can influence many profoundly important choices we are forced to make, from the food we eat to the forms of energy we use.

Our society has a complicated relationship with science. On the one hand, we have entered a post-truth era in which politicians publicly speak of alternative facts when actual facts are inconvenient. On the other hand, "follow the science" has become a mantra. Having a clear and compelling account of what science can achieve could help us navigate between an ignorant dismissal of science and a naive trust in science. Ultimately, trust in science should be based on the account of deontic evaluation I described in chapter 2, itself based on the extent of satisfaction of justificatory norms described in chapter 1. That some scientific claim is a timeless truth is not a necessary condition for trusting or relying on that claim. Yet defending the possibility of timeless truths is a helpful tonic, considering the volume of scholarly work about science that appears to completely dismiss the possibility of timeless truths by claiming that all scientific findings are provisional.

The pessimistic meta-induction is a famous argument suggesting that all scientific theories from the past have eventually been proven false, and there is no reason to believe that our current scientific theories are any more accurate than those of the past, thus all of our current scientific theories are very likely to be proven false in the future (Laudan 1981). Examples of scientific theories or entities that were once popular but have since been rejected include Ptolemaic astronomy, phlogiston, ether, and caloric.

Although the pessimistic meta-induction has been a staple argument for antirealists and could also be an argument for the provisionality thesis, Vickers (2023) rightly argues that the realism-antirealism debate is orthogonal to the provisionality thesis. That is because even antirealists articulate conditions under which science can attain truth. Van Fraassen (1980), for example, argues that science can attain knowledge of the observable realm, and we should be antirealists about unobservables. Leaving aside the problem of explaining what exactly *observable* and *unobservable* mean, according to van Fraassen's view, science can attain truths about at least some things. He illustrates the point with flair (15): "A flying horse is observable—that is why we are so sure that there aren't any." There is no reason to think such truths are not timeless. So, antirealists do not need to agree with the provisionality thesis, and realists should actively reject it. To be clear, the aim of this chapter is not to enter the old debate between realists and antirealists. Rather, I argue for the more humble thesis that some claims of science, namely, elements of common knowledge, are timeless truths or, to use Vickers's term, future-proof. While this thesis of timeless truths may seem to be closely aligned with scientific realism, it is a more modest position, one that should be agreeable to both scientific realists and scientific antirealists (§8.4).

This chapter differs from the previous chapters in one central way: It emphasizes science's ability to achieve its constitutive aim, while the other chapters emphasize the quality of the pursuit rather than the achievement of the aim. Demonstrating this is important because the provisionality thesis is so often asserted in discussions of science. I argued earlier that the attainment of truth and knowledge should not be the basis of evaluative concepts for science. Yet that does not entail that truth or knowledge is either unimportant or unattainable. I argued in chapter 1 that the constitutive aim of science is a special kind of knowledge, common knowledge. Science can attain common knowledge. There is no contradiction in maintaining that the aim of science is a special kind of knowledge while also maintaining that the attainment of knowledge should not be the standard for evaluative concepts in science, because, as I argued in the introduction and chapter 1, we should be nonconsequentialists about our evaluative standards.

I begin this chapter with a brief background of the provisionality thesis, speculating why it has become such a widely stated platitude (§8.2). I then argue against the provisionality thesis, arguing that some scientific findings are timeless truths—some scientific findings will not, even in the distant future, be overturned (§8.3). Part of the argument for timeless truths involves the concession of antirealists that science can attain timeless truths as long as what I call a "knowledge-possible" condition is satisfied. In §8.4, I briefly explore some proposed knowledge-possible conditions. Sometimes, the seductively intuitive idea that science is a never-ending story—that there will always be new things to learn about the universe, perhaps at more extreme spatiotemporal scales—gets mixed up with the provisionality thesis. In §8.5, I note that an argument in favor of the never-ending story is not an argument for the provisionality thesis; they are theses about completely different aspects of science. The provisionality thesis says that any particular scientific theory is provisional, while the never-ending story says that even if particular scientific theories about particular domains are timeless truths, there will still be unanswered questions about other domains or at different spatiotemporal scales. I address several objections in §8.6 before concluding in §8.7.

8.2 A Battle for Final Truth

Almost certainly, Karl Popper is the most famous philosopher of science among scientists themselves. The main idea of Popper's (1963) philosophy of science was that scientific theories never receive any confirmation, they cannot be deemed true, and they should be considered simply as provisional hypotheses. This view of science is called *falsificationism*, as it says that good science involves scientists formulating a risky theory for which evidence could be acquired that would falsify that theory, and scientists should do all they can to try to falsify their theories. The theories that (temporarily) survive this rigorous process should be thought of as provisional conjectures, and no amount of supporting evidence can change our assessment of the probable truth of a theory (for problems with this view, see chapter 2).

Perhaps the second-most famous philosopher of science among scientists is Thomas Kuhn. The main idea of Kuhn's (1962) philosophy of science is that science undergoes occasional "revolutions" in which an existing paradigm gets overturned and replaced with a new paradigm. For example, the Ptolemaic model of the solar system placed the earth in the center, and that was replaced by the Copernican model, which placed the sun in the center. Kuhn suggested that such revolutions would continually cycle in science. If that is true, currently endorsed scientific theories will also be replaced, and thus our attitude toward them should be provisional. Laudan (1981) developed this consideration into the pessimistic meta-induction.

A feature of science that may support the belief in the provisionality thesis is the importance of organized skepticism and criticism in science, noted by Merton ([1942] 1973) and more recently emphasized by Longino (1990). Scientists are trained to be epistemically humble and to view the results of their work as fallible (Elgin 2017, 111). Epistemic humility is a noble trait for people in general, and it is especially important for scientists. One might think that a principle of epistemic humility suggests that scientists should deem their conclusions as provisional.

The impact of Popper and Kuhn, and the intuitive appeal of the Merton–Longino norms of organized skepticism, criticism, and humility, perhaps contribute to the putative attraction of the provisionality thesis. A wide range of philosophers, historians, and sociologists of science assert the provisionality thesis. Longino (2002, 9), for example, explicitly affirms the provisionality thesis. Stanford (2015, 875) claims that our "theoretical conceptions of nature will continue to change just as profoundly and fundamentally as they have in the past." Elgin (2017) emphasized "the permanent possibility of error" (6) and thus states that "science does not consider any methods or results incontrovertible" (103). Collins and Evans (2017, 19) claim that "science's understandings are continually disputed."

Wray (2018, 2) asserts that we have "reason to believe that many of our best theories are apt to be rendered obsolete in the future." Cartwright et al. (2022) agree: "There is a good reason to expect the overthrow of 'well-established' scientific principles." Potochnik (2017, 121) also agrees: "Our current best theories are probably not true." Mitchell (2020, 790) hints that it is "fictional" to claim that science can attain "eternal truth." Oreskes (2019, 49), in her book unironically titled *Why Trust Science?*, claims that "the contributions of science cannot be viewed as permanent. The empirical evidence gleaned from the history of science shows that scientific truths are perishable." Yet it is hard to see why one should rely on the products of an endeavor if those products are perishable.

Many scientists appear to believe that science achieves timeless truths (see Beebe and Dellsén [2020] for a survey). Occasionally, scientists make claims about science that seem to contradict the provisionality thesis and claim that science can attain timeless truths. For example, as we saw in the introduction, physicist Lise Meitner, who helped discover nuclear fission, claimed that science is "a battle for final truth" (Rhodes 1986, 234). Who is right?

8.3 Timeless Truths

Vickers (2023) argues, convincingly in my view, that some scientific findings are "future-proof," which means that the science of the future, even the distant future, will continue to uphold some scientific findings of today. Consider some examples. One thousand years from now, scientists will continue to believe that water is H_2O, that the structure of DNA is a double helix, that the tides are caused by the moon's gravity, and that uranium atoms can be split into smaller atoms and thereby release a vast amount of energy. The many claims among philosophers of science, historians of science, and scientists themselves in support of the provisionality thesis, some of which I noted in §8.2, are, thus, misguided.

Consider an example from metrology. John Norton noted that estimates of the charge and mass of the electron have remained virtually unchanged for over eighty years. Since he wrote this in 2000, we can say that our estimates of the charge and mass of the electron have stayed nearly the same for over a century. Since 1940, the measured charge of the electron has not changed out to twenty-two decimal places of a coulomb, with many research groups using a diverse range of ever-improving measurement techniques over these decades. A typical smartphone has about ten thousand coulombs in its battery. To visualize the precision and stability of the measurement of the electron charge, imagine taking a smartphone battery and crushing it into ten thousand grains of sand, then taking one of those grains of sand and crushing it into one billion invisible specks of dust, then taking one of those specks of dust and crushing it into another one billion infinitesimal flecks, then taking one of those infinitesimal flecks and measuring its charge to *four* decimal places of a coulomb. In the entire century of measuring the charge of that infinitesimal fleck, the values of our measurements would not have changed, despite all of the advances in technology. So, the measurement of the electron charge has for a very long time been extremely precise and stable. Of course, measurements of the electron charge will continue to get more sophisticated, and thus there will be changes to our estimates at, say, the thirtieth decimal of a coulomb, but that is irrelevant to the provisionality thesis, as it is a version of the never-ending story (discussed below). If we consider a suitably hedged estimate of the charge of the electron (in the sense introduced in chapters 3 and 4) by considering the charge of the electron at twenty decimal places of a coulomb, that value is a timeless truth. The measured charge of the electron now is $1.60217663 \times 10^{-19}$ coulombs, and so the measured charge of the electron out to the twentieth decimal place of a coulomb is 1.6, and that is definitely a timeless truth.

Vickers focuses primarily on articulating and defending a set of criteria that can be used by nonexperts to identify future-proof science. These criteria are elegantly simple: 95 percent of a relevant scientific community must believe a scientific claim (the *consensus* condition) and that scientific community must be sufficiently diverse (the *diversity* condition). Vickers draws on a range of examples to argue for the following cluster of claims that hang together in support of the future-proof thesis:

(i) Even antirealists agree that science can discover truths in some circumstances, namely, if we are on the right side of whatever knowledge-possible condition they adhere to, such as observability (§8.4); such truths are timeless.
(ii) Contrary to the pessimistic meta-induction, we learn from Fahrbach (2011) and others that science is not a graveyard of dead theories, and thus the pessimistic meta-induction is not as threatening to the future-proof thesis as antirealists often claim (I return to this point below).
(iii) Whether a scientific claim is future-proof is fundamentally determined by the first-order evidence for that claim (more on this below).

(iv) Nonexperts are better off looking not to the first-order evidence for a scientific theory to assess whether that theory is future-proof but to the consensus and diversity conditions because those higher-order conditions are easier to assess for nonexperts than is the vast and complex first-order evidence.
(v) The inference from the satisfaction of the consensus and diversity conditions to a claim about future-proofness has a stellar track record (100 percent!)—that is, for every instance in which the consensus and diversity conditions have been satisfied, the scientific theory in question has withstood the test of time, has not been overturned, and is now aptly deemed future-proof.
(vi) Some intuitive examples of future-proof scientific theories now satisfy the consensus and diversity conditions, like the examples in the first paragraph of this section (water is H_2O, the structure of DNA is a double helix, the uranium atom can be split, and so on).
(vii) Putative historical counterexamples—examples of scientific theories that seem as if they satisfied the consensus and diversity conditions yet later came to be overturned—can be dispelled by showing that in fact one or both of the consensus and diversity conditions were not satisfied.

Vickers devotes chapters with detailed case studies to defend each of (i)–(vii). I will not reiterate the details of those arguments, as readers can consult Vickers's book. I aim to articulate what I consider the best positive argument for the future-proof thesis. For Vickers (2023, 19), that argument is based on "the quantity and the quality of the evidence for these ideas, vetted by thousands of scientists, embedded within a sufficiently diverse scientific community." Vickers notes the importance not merely of consensus for future-proof science but also of "hard won" consensus. The notion of common knowledge, described in chapter 1, can support Vickers's defense of future-proof science.

To challenge the provisionality thesis and defend the existence of timeless truths, one must address the historical basis of the pessimistic meta-induction since it seems to have captured the imagination of many scholars. The assertion of the provisionality thesis is usually paired with a brief remark that the history of science is a graveyard of dead theories. Yet critics of the pessimistic meta-induction note that the historical evidence presented by Kuhn and his followers for the provisionality thesis and the examples used by Laudan to support the pessimistic meta-induction appear cherry-picked. Kuhn's examples concern theories with a high degree of empirical underdetermination, such as models of the solar system until the seventeenth century. Kitcher (1993) argues that the history of science offers grounds for optimism rather than pessimism, at least for some scientific claims. Fahrbach (2011) gives a compelling argument that science today is different in many respects from sciences of the past. The idea is that the quality and quantity of scientific work increases exponentially over time, and the reliability and sensitivity of new instruments and computing power has also dramatically increased. This entails that the present is quite different from the past regarding science's capacity to discover timeless truths; thus, the overturning of past scientific findings does not give us reason to expect future overturning of present scientific findings (see also Mizrahi 2016). Following Fahrbach, Bird is impressed that of the hundreds of Nobel Prizes awarded in the twentieth century in physics, chemistry, and physiology/medicine, the scientific work that formed the basis for the prizes was subsequently overturned only once. Thus the primary argument supporting the provisionality thesis, and thus one thought to raise doubt for the existence of timeless truths, has, to use Bird's (2022, 44) words, "run out of steam."

In short, Vickers, Fahrbach, Mizrahi, and Bird resist the pessimistic meta-induction by noting its weak empirical basis (see also Magnus and Callender [2004], who challenge the formal aspect of the pessimistic meta-induction). Lipton (2004, 1266) makes a slightly different complaint about the pessimistic meta-induction when he calls it "judo epistemology": "The strength of current science—its improvement over past science—is thus used against itself," which he rightly considers suspicious. The pessimistic meta-induction has indeed run out of steam and thus should not be seen as a barrier to the possibility of timeless truths. In the remainder of this section, I offer a positive argument for timeless truths, which proceeds in three steps.

First, antirealists grant that we can put aside evil demon scenarios. An evil demon scenario is one in which all our experiences of the world are misleading because we are being systematically tricked by an evil demon, or we are brains in vats plugged into a powerful computer. Some philosophers maintain that there is a nonnegligible possibility that our lives and experiences are just simulations in a powerful computer (Chalmers 2022). If we are in an evil demon scenario, then the provisionality thesis follows trivially since the computer programmer who is simulating our experience of the world could change the laws of the simulated universe today. If this happened, our science of tomorrow would have to correspondingly change, and scientific theories of today would be discarded. Similarly, if the laboratory staff managing the matrix that we are plugged into could disconnect us and introduce us to the real world, and if the real world were nothing like the matrix world, our scientific findings that appeared future-proof in the matrix world would be discarded, and, again, the provisionality thesis would follow trivially. These considerations should give little consolation to the defenders of the provisionality thesis. They should first ask how likely it is that we are in an evil demon scenario. Even if we are in such a scenario, the timeless truths versus provisionality debate arises within that scenario. We have no reason to think that our simulators are changing the laws of the simulated universe or somehow fooling us into thinking that simulated DNA has a double-helical structure—all of our evidence suggests otherwise, and so, as good empiricists, we can conclude that the universe, simulated or not, has various features. Supporters of the timeless truths thesis say we can have timeless truths about these features, and supporters of the provisionality thesis say we cannot.

In any case, antirealists tend not to be global skeptics about knowledge. Antirealists agree that we can attain knowledge of the world as long as particular conditions are satisfied—I call such conditions "knowledge-possible" conditions. Perhaps the best-known knowledge-possible condition is "observability": If x is observable, some antirealists claim, we can have knowledge about x, and if x is unobservable, we cannot have knowledge about x (van Fraassen 1980). There is no reason to think that antirealists would deny that such knowledge is timeless. Thus, even antirealists should accept that there can be timeless truths. Much depends, however, on the knowledge-possible condition. If a knowledge-possible condition is stringent—say, if one adopts the observability condition and if "observability" requires actual capacity to observe an entity or process with one's own eyes, unaided by technology—then most scientific claims will not satisfy the knowledge-possible condition, and our timeless truths will be limited to claims like "the sky is blue." On the other hand, if a knowledge-possible condition is relaxed—say, a "some evidence" condition that says knowledge about x is possible if at least some evidence can be gathered for x—then all sorts of scientific, nonscientific, and even pseudoscientific claims could satisfy the condition, and our stock of putative timeless truths would include claims that should be deemed neither future-proof nor knowledge. This suggests that a knowledge-possible condition should be less stringent than the extreme form of strict unaided observability and more stringent than a "some evidence" condition. In §8.4, I discuss some proposed knowledge-possible conditions and argue for what I consider to be the most compelling knowledge-possible condition based on the

Leibniz procedure from chapter 1. Since science at its best can satisfy this condition, science can attain timeless truths.

This leads to the second step in my argument for timeless truths. In the first step, I noted that even antirealists must grant that we can attain timeless truths as long as a knowledge-possible condition is satisfied. The second step notes that the boundary of applying any plausible knowledge-possible condition expands over time. Science involves improvements in instruments and justificatory practices, and this entails that the domain in which a knowledge-possible condition can be satisfied continuously expands. The spatiotemporal domain of what is "observable" increases with innovations in instruments. This is a very strong argument for timeless truths. Vickers (2023, 124) calls an implication of this expansion "test by observation," which involves the technology-aided observation of entities and processes that were not previously observable. He notes that we can now take photographs of entities at a scale that would have been unthinkable a few decades ago, such as the double-helical structure of DNA imaged by transmission electron microscopy (203). So the development of novel instruments expands the scale at which knowledge is possible (see §8.4 below) and also affords very stringent empirical tests of the timeless truths thesis.

The third step in the argument for timeless truths relies on the view that I defended in chapter 1: A constitutive aim of science is "common knowledge," which is not mere consensus; it is strong consensus, one in which all parties to the consensus accept each other's justifications for the claim about which there is consensus. We saw earlier that this aim entails a kind of Merton-style communism and the importance of Longino-style criticism for science, the result of which is a very robust justificatory basis for claims about which there is strong consensus. Massimi (2022a, 5) correctly notes that the reliability of science is a result of the "deeply *social* and *cooperative* nature of scientific inquiry." This special reliability of science contributes to the capacity of science to attain timeless truths. The many justificatory principles and practices in science and their rapid improvement over time in the context of the social and cooperative (and critical) nature of science, entail that science can attain timeless truths, or common knowledge. Though this is a departure from Vickers's argument for future-proof science, it is consistent with it. As noted above, Vickers claims that the fundamental basis for future-proof science is the first-order empirical evidence for scientific findings, and the epistemological structure of the Leibniz procedure explains how first-order empirical evidence can be so inductively powerful. Common knowledge requires meeting the first-order epistemic criterion for future-proofness and requires that the consensus and diversity conditions for identifying future-proof science that Vickers defends are manifest and will likely be ascertainable. The two foundational notions described in chapter 1—common knowledge and the Leibniz procedure—provide bedrock for the argument that science can attain timeless truths.

8.4 Knowledge-Possible Conditions

As mentioned, even antirealists grant that under particular circumstances, we can have access to timeless truths. This is exemplified by van Fraassen, whose knowledge-possible condition is observability, and Wray (2018), who also seems to adopt this knowledge-possible condition.

Sounding surprisingly similar to the logical empiricists of the last century, Chang (2022, 6) states that it is impossible for science to "attain assured truths about what truly goes beyond experience." So, on this old-fashioned view, a knowledge-possible condition for a claim x would be something like "ability to experience x." There is one interpretation of this thesis that is implausible and another that is unthreatening for the timeless truths thesis. The implausible interpretation is that experience is to be understood as perceptual evidence that humans can directly attain, such as my current experience of a coffee cup on my desk. If that is what Chang means by experience, then the thesis that science cannot attain truths that go beyond experience is implausible because science devises ingenious methods to expand the domain in which we can attain truth-conducive evidence (this interpretation would rely on a crude empiricism that I would expect Chang to reject; see, for example, Chang [2022], 61). On the other hand, if one interprets "beyond experience" as something like "beyond any possible evidence whatsoever," then the thesis is plausible—of course, science cannot attain truths based on no evidence. But this is irrelevant to most science and would not challenge the timeless truths thesis because science obviously produces a wealth of diverse and compelling evidence for many theories (Vickers [2020] makes a similar point to address Wray's [2018] notion of "unobservable").

A better knowledge-possible condition is suggested by Hoefer (2020), who argues that because fundamental physics, specifically quantum mechanics, is particularly vulnerable to underdetermination, contemporary claims in fundamental physics should be considered as provisional, though other areas of science have theories that are candidate claims to knowledge. Hoefer's knowledge-possible condition is something like: x can be a candidate for knowledge if and only if x is not in the domain of fundamental physics (see also Callender [2020]; for criticism of Hoefer's account, see Vickers [2023, 142–44]). Hoefer's (2020, 34) argument relies on what he calls "epistemic handles," meaning "direct or indirect observations; theoretical reasons for believing that an entity of such a type should exist; knowing how to produce and manipulate the entity; taking the entity to be the cause of a certain observable event or phenomenon (where this is not a case of observation), etc." The concept of an epistemic handle can be grounded by my account of justificatory norms developed in chapter 1. Hoefer notes that insufficient epistemic handles exist in fundamental physics to ensure that theories are true, but in other scientific domains, sufficient epistemic handles exist to ensure that at least some theories in those domains are true and thus not liable to be overturned (though it is plausible that the availability of epistemic handles is theory-relative rather than domain-relative).

A more general knowledge-possible condition, one that undergirds those already mentioned, is based on the spatiotemporal scale of the objects of study. Objects that exist at extremely small scales, such as the postulated entities of fundamental physics, are difficult to get epistemic handles on. The same is true of objects at great distances from us or of phenomena at extreme timescales, such as the structure of the universe nanoseconds after the Big Bang. Crucially, science's capacity to provide rigorous justification depends on the spatiotemporal scale of the object of study. For objects of study at very large scales (like properties of the distant universe) and very small scales (like the smallest constituents of matter), our epistemic limitations are so severe that we may never be able to provide much justification for claims about objects at these extreme scales. Therefore, our hypotheses about the universe at such scales will likely always be provisional. But for many objects of study at middling scales (say, from basic features of nuclear physics up to the structure of our galaxy), we can provide incredibly rigorous justification for many hypotheses. A simple and extreme example is "there is a cup of tea on my desk," which is both true and maximally justified and is provisional only in the sense that locations of cups of tea generally are.

Commenting on the role of scale in determining a knowledge-possible condition, Vickers (2023, 145) claims that because there are things we can know at some very small scales, a knowledge-possible condition should not be defined by reference to scale. Vickers instead proposes the notion of a "concept application problem": In some domains, particularly the quantum domain, it is awkward or impossible to apply concepts that we successfully use in more familiar domains. And when a concept application problem arises, our claims to knowledge are plausibly provisional. While I find the notion of a concept application problem intriguing, Vickers notes that concept application problems arise prominently at extreme scales; his example is, like Hoefer's, quantum mechanics. Vickers is correct that we can know things at very small scales—we have known for more than one hundred years, for example, that atoms are composed of a dense nuclear core. These examples do not show, however, that scale is not the fundamental basis of a knowledge-possible condition. Rather, such examples show that scientists can develop ingenious methods to expand the scale at which knowledge is possible. The simple, elegant gold foil experiments of Hans Geiger and Ernest Marsden, around 1909, were enough for Ernest Rutherford to conclude that atoms have a dense nuclear core, thereby expanding the boundary of the scale at which knowledge is possible.

In chapter 1, I argued that common knowledge requires a strong consensus, and that in turn requires higher-order deliberation and consensus about justificatory norms. I introduced the Leibniz procedure as an epistemological device to undergird the normative basis of justificatory norms and their relationship to first-order scientific claims. The requirements of common knowledge, attained by satisfying the deliverances of the Leibniz procedure, are very demanding. Yet science can attain knowledge by satisfying its many justificatory norms, and such knowledge can ultimately achieve the status of common knowledge. Common knowledge is composed of timeless truths.

8.5 The Never-Ending Story

Defenders of provisionality sometimes seem to conflate *timeless truths* with *finality*, the view that science can or will discover all truths, or all truths about a particular object, phenomenon, or domain, and thereby come to an end. The timeless truths thesis is about specific scientific claims, including singular facts and general theories. The fact that there is always more to learn about the world does not speak against the possibility of timeless truths. Science might be a never-ending story insofar as there might always be new aspects of nature to discover, particularly at more extreme spatiotemporal scales, but that does not entail that particular discoveries are necessarily provisional. So, you could adopt the never-ending story view of science, but this should not lead you to infer the provisionality thesis.

As an example of what appears to be such an inference, Elgin (2017, 60) writes that "Copernicus's central claim was strictly false" on the grounds that planets do not travel around the sun in a circular orbit. Yet articulated at a coarser grain, one more appropriate for describing Copernicus's achievement, the "central claim" was that the sun is the center of the solar system, which, of course, is true. A fussy interlocutor might say that the center of mass of the solar system is not the sun itself but is near the sun, though that pedanticism is irrelevant to the core truth of Copernicus's heliocentric model, which is simply that the planets orbit around the sun. When stated at a particular grain of detail, Copernicus's central claim is clearly a timeless truth. What Elgin rightly emphasizes is that, despite Copernicus's great breakthrough, there was still more to learn about the structure of the solar system and the motion of planets. Kepler proved that planets move in elliptical orbits; Newton taught us that gravitational forces between planets cause their orbits to be slightly nonelliptical; today we know that the sun also moves in response to the gravitational forces of the planets, and the entire solar system is moving at a whopping 250 kilometers per second relative to the center of the Milky Way. There is still more to learn—in Elgin's terms (2017, 60), "no one claims that science has as yet arrived at the ultimate truth about the motion of the planets." The word *ultimate* here is ambiguous, though. It could be understood as "final," in contrast with the never-ending story, in which case, Elgin's claim would, for now at least, be apt. Yet it could also be understood as "timeless," in contrast with provisional, in which case, Elgin's claim would be misleading since the planets really do orbit around the earth. Though our study of planetary and solar motion may be a never-ending story, a description of the solar system's structure, described at the appropriate level of detail to avoid pedantic fussiness, is a timeless truth.

Massimi (2022a, 20) argues for the never-ending story based on her thesis that scientific knowledge is perspectival: "The absence of a scientific 'view from nowhere' means there exists no ideal atlas, and no privileged catalogue of ontological units. There is no directionality to serve in worn-out discussions about convergence towards a final theory or a final metaphysical reality or end of inquiry. Nor one that is synonymous with consensus building as a way of homogenizing lines of inquiry and smoothing out pluralism. There is, in other words, no conclusion to the story of our scientific endeavours. Scientific knowledge is ongoing and open-ended." However, one can grant that science is a never-ending story while also maintaining that particular scientific findings are timeless truths, as with the example of the relative position of the planets to the sun. Moreover, candidate timeless truths have typically been examined from diverse perspectives (the epistemic importance of which Massimi emphasizes), which contributes to their status as candidate claims to common knowledge. The claim that scientific knowledge "is ongoing and open-ended" is misleading—science itself may be ongoing, in the never-ending story view of science, but that does not mean that particular claims to scientific knowledge are not timeless. To think so would be to commit the fallacy of inferring from the never-ending story view of science to the provisionality thesis.

One might insist that specific scientific results must be provisional because the technologies that produce these results are routinely changing and improving, leading to changes in the results themselves. Measurements of physical constants provide a simple illustration: New and better measurements of physical constants are made on a routine basis, and there is no reason to suppose that such developments will not continue; therefore, all such results are provisional. Yet this line of reasoning is another example of the fallacy of inferring provisionality from the never-ending story. Consider again the measurement of the charge of the electron (§8.3). While it is almost certainly true that future theoretical and technological developments will enable us to measure the charge of the electron out to more and more decimal places of a coulomb, those future measurements will *confirm* the existing measurements of the electron charge at, say, twenty-one decimal places of a coulomb. We can be extremely confident of this because the many diverse methods of measuring the electron charge over the last eighty years have all agreed on its value at twenty-two decimal places of a coulomb, and measurements in the last fifty years all agree on the value at twenty-three decimal places of a coulomb, so the value of the electron charge at twenty-one decimal places is a timeless truth. de Melo-Martín (2024, 7) claims that Vickers's examples of future-proof scientific findings are "vague or too broad," but there is nothing vague or too broad about the electron charge at twenty-one decimal places of a coulomb. On the contrary, this is a scientific finding of great specificity and precision.

8.6 A Place Beyond Heaven

Some might argue that the denial of the provisionality thesis is a display of hubris. Does the affirmation of timeless truths violate a principle of epistemic humility? Epistemic humility for scientists pertains primarily to their research at the cutting edge of inquiry. For a particular scientist, it is reasonable to believe that her recent research findings are provisional; throughout this book we have seen many examples of great scientific discoveries being presented as provisional, as with Avery's discovery that genes are composed of nucleic acids, Watson and Crick's discovery of the double-helical structure of DNA, and Chadwick's demonstration of the existence of neutrons. Yet it is equally reasonable for a scientist to believe that some of the best-established findings in their discipline—not at the cutting edge of research—are not provisional but rather are timeless truths.

In a passage apparently objecting to the possibility of timeless truths, Chang (2022, 104) claims that the appeal to basic examples of timeless truths such as the chemical structure of water and the causes of tides "begs the questions of why we feel so certain about such propositions, and whether we are justified in that certainty." He proceeds to note that much of his previous work "has been devoted to showing that the establishment of such scientific truths is a long and painstaking process" (104). Yet one can grant that the establishment of scientific findings involves a long and painstaking process while also admitting that certainty in some findings can be justified. Indeed, the difficulty of achieving common knowledge is what ultimately warrants timeless truths. In chapter 1, I quoted Marie Curie, who claimed, "I was taught that the way of progress was neither swift nor easy." Thanks to Marie Curie's heroic efforts, our knowledge of many aspects of radiation and the elements radium and polonium are now part of the public treasury of scientific common knowledge. The difficulty of achieving common knowledge is not an argument against the possibility of timeless truths but rather an argument in its favor.

There is an old tradition in philosophy of characterizing aspects of the world as being in-principle inaccessible by human experience. From Plato's forms to Kant's noumena, philosophers have long postulated deep, dark, inaccessible aspects of nature, hidden in a "place beyond heaven" as Plato puts it in his dialogue *Phaedrus*. Philosophers of science sometimes adopt and develop this tradition, declaring that all human knowledge is perspectival (Massimi 2022a) or that scientific knowledge is necessarily "mind-framed" (Chang 2022), and thus there cannot be timeless truths. As Chang (2022, 6) puts it, "It is time to accept the fact that we cannot know whether we have got the Truth about the World." Yet Chang presumably believes it is a truth—with a capital *T*, whatever that is supposed to signify—that, say, the planets orbit the sun. This Platonic tradition involves a profound mistake, premised on the idea of a place beyond heaven. As Kitcher (1993, 118) elegantly puts it, scholars today who continue this tradition demonstrate not a metaphysical thesis but rather a thesis about the scholars themselves: "Reticence about the attainment of significant truth is the result of a failure of nerve." Along the line of Fine's natural ontological attitude, determining whether there are deep, dark, hidden parts of nature should be based on the results of the scientific study of nature itself rather than a philosopher's lack of nerve.

Above I noted that a plausible knowledge-possible condition is based on scale. Objects at extreme physical scales may be sufficiently inaccessible such that we cannot have secure knowledge of those objects. But for objects on our side of the knowledge-possible condition—that is, objects at scales such that we can develop sufficient epistemic handles on those objects—science can have knowledge of such objects. Some parts of nature are hidden from us just as if we were stuck in Plato's cave, but other parts of nature are not hidden from us—and, as argued above, the domain that is accessible to us grows rapidly thanks to technological developments—and thus we regularly emerge from Plato's cave to observe nature as it is.

Consider a science of aliens, or the science of artificial intelligence in the not-so-distant future, or science from the far future, say, one million years from now. Can we even imagine what such science would look like? Can we assume that the claims of such unfamiliar sciences would be anything like the claims of our science today? One might suspect that such sciences would contradict our current claims to common knowledge. Yet this conclusion would be hasty. For contributions to common knowledge, anything other than a positive answer to those questions would need a compelling independent argument. I see no reason to think that when aliens finally visit us and study our world, they would find that the structure of water is something other than H_2O or that atoms do not have a dense nuclear core or that the sun is not in the center of the solar system. On the contrary, when aliens visit us and, say, study the structure of DNA, we have every reason to think that they will conclude that it is a double helix.

Of course, the science of aliens or future artificial intelligence would very likely use different concepts and categories than our science and would study aspects of nature that are salient to them and which may not be so to us, and they could possibly observe nature at more extreme physical scales than we can. But if we agreed with them on a translation between concepts and asked them to study the macrostructure of DNA, those alien or artificially intelligent scientists, if they were sufficiently competent, would conclude that the structure of DNA is a double helix. Obviously, they might get it wrong, but if they did, we would be able to describe flaws in their putative justification for their conclusion, and though the expanded community of science (us and them) would temporarily revert to a pre-common knowledge state about this claim (since we would no longer have consensus about the claim and would disagree about our respective justifications), the process described in chapter 1 of articulating and criticizing putative justifications and revising and improving those justifications would ultimately bring that expanded community back to the original contribution to common knowledge, namely, that the structure of DNA is a double helix. It is not arrogant to be certain that it would be their justification and conclusion and not ours that would need revising because our justification and conclusion have already satisfied the extremely demanding requirements of common knowledge.

Commenting on Vickers's future-proof thesis, de Melo-Martín (2024, 9) suggests that there would be no practical difference for people if they knew that a particular scientific claim is a timeless truth. She asks, rhetorically, "What difference would it make to people's actions to determine that a particular scientific claim is future proof?" De Melo-Martín claims that for many of the examples that Vickers gives of future-proof findings, the practical consequences for people believing that those findings are future-proof are not clear, and she mentions the example that the sun is a star to suggest that the status of a claim as a timeless truth is "unlikely to do much" (9).

To answer de Melo-Martín's rhetorical question, I say: All the difference in the world. If a scientific finding is indeed a contribution to common knowledge, it becomes part of Leibniz's public treasury (chapter 1), and so people can adopt the finding as given in their practical deliberations and safely assume that others do so too, thereby aiding social coordination. The practical consequences of particular findings becoming common knowledge can be extremely impactful, and we may not always notice such consequences precisely because those consequences have been so profoundly consequential: Belief that the sun is a star precludes belief that the sun is a god; widespread acceptance of the germ theory of disease motivates broad use of hygienic measures such as handwashing or public health measures such as vaccination; believing that the sun and not the earth is at the center of the solar system involved a massive reconceptualization of our place in the universe; the nineteenth-century study of stratigraphy showed that the earth is vastly older than previously thought, and this finding, now common knowledge, had deep implications for religious and mythical beliefs and our subsequent tenuous grasp on the meaning of life itself.

The profound consequences of these timeless truths include a self-reinforcing philosophical implication: Having learned that there is no heaven, we learn that there is no place beyond heaven. This renders the Platonic tradition even more dubious, and since it is this tradition that philosophers of science often draw on when promulgating the provisionality thesis, learning that there is no place beyond heaven boosts our confidence in our capacity to attain timeless truths.

Denying the possibility of science attaining timeless truths—as many thinkers do when they assert the provisionality thesis—could contribute to general distrust in science, whereas to affirm the actual achievement of some timeless truths would be to celebrate one of the most important accomplishments of science, which could plausibly contribute to increasing trust in science's general ability to teach us about the world. This is another way in which the affirmation of timeless truths is far from idle.

The argument in this chapter might seem discordant with the previous chapters since I have spent most of this book arguing that our evaluative concepts for science should be based on justification rather than truth or knowledge—yet, here I am arguing that science can attain timeless truths. Recall, however, that in the introduction, I defended nonconsequentialism about goal-seeking: It is perfectly coherent to argue that the constitutive aim of science is a special kind of truth, namely, common knowledge, and that science can achieve this aim while also maintaining that our evaluative concepts for science should be based not on the attainment of its aim but on the quality of the means taken toward that aim. The slogan was: Good science need not attain its aims, it must justify its claims. Yet that does not imply that science cannot attain its constitutive aim. In this chapter, I have argued that it can.

8.7 Conclusion

To determine whether a scientific finding is provisional or is rather a timeless truth is to make an evaluative judgment. As I argued in chapter 2, our evaluative judgments of science should not be conducted in a coarse-grained way that categorizes all of science as being on one side or the other of an evaluative judgment. If all elements in some domain are judged to be the same based on some evaluative concept, then that is hardly an evaluative concept worth having. Imagine, for example, the Cambridge Academy of Failure, in which all pupils fail their course of study and have done so for the entire thousand-year history of the academy. Perhaps the pupils fail because they must fail—it is simply a feature of the difficulty of the curriculum that they cannot pass. Or perhaps we only observe that pupils have always failed, so we conclude that all current pupils are also bound to fail. At some point, it would be worth asking, What is the purpose of this evaluative scheme? Since all students fail, the scheme is not doing any real work. Promoters of the provisionality thesis argue that when assessing whether a scientific finding is provisional or a timeless truth, all scientific findings are like the pupils at the Cambridge Academy of Failure—they are all liable to fail. This view of science should be rejected.

It is possible that the many scholars I cited at the beginning of this chapter do not really believe the provisionality thesis in any general sense. Perhaps the provisionality thesis has just become a manner of expression, or perhaps its plausibility was exaggerated because of the focus in twentieth-century philosophy of science on theoretical physics, the domain of science in which provisionality is most plausible. Those scholars who express the provisionality thesis, when pressed, may agree that dinosaurs existed, that water is H_2O, and that the planets orbit the sun—at least, I hope they would. Yet the many assertions of the provisionality thesis suggest that it is taken seriously as a view in philosophy, history, and sociology of science, and scientists themselves sometimes express the view. I hope the arguments in this chapter help to demonstrate the implausibility of the provisionality thesis and, in turn, the plausibility of timeless truths.

Putnam ([1974] 1991, 122–23) colorfully noted that the distinction between provisional conjectures and timeless contributions to knowledge is important. Our judgment that a scientific finding is either provisional or a timeless truth carries weight in the world: "The advice to regard all knowledge as 'provisional conjectures' is also not reasonable. Consider men striking against sweat-shop conditions. Should they say 'it is only a provisional conjecture that the boss is a bastard. Let us call off our strike and try appealing to his better nature.' The distinction between knowledge and conjecture does real work in our lives." So the distinction between provisional conjectures and timeless truths is important, and that would only be the case if elements on either side of the distinction existed not in a place beyond heaven but right here on Earth. The heart of science is justification, and the strict demands of scientific justification—the demands of the Leibniz procedure—create the conditions under which the constitutive aim of science can sometimes be achieved. The aim of science, sometimes achieved after hard work, is common knowledge, and contributions to common knowledge are timeless truths.

Conclusion

In the introduction, I developed the notion of a deontic philosophy of science and argued for its merits. I developed the foundation of this deontic philosophy of science in chapter 1, arguing that the constitutive aim of science is common knowledge and that justificatory norms for science get their normative status via the epistemological device I dubbed the Leibniz procedure. I then used this foundation to offer novel accounts of a range of philosophical notions for science, including scientific evaluation (chapter 2), the value-free ideal (chapter 3), scientific assertion (chapter 4), scientific progress (chapter 5), scientific credit (chapter 6), and fast science (chapter 7). I closed in chapter 8 by arguing that science can attain timeless truths, contributions to common knowledge that are secure and not provisional.

I hope that the deontic philosophy of science developed throughout this book will illuminate a broad range of topics in philosophy of science, beyond those addressed in the prior chapters. Some possibilities include a novel account of scientific explanation with an emphasis on justificatory norms; a novel account of pursuit-worthiness given in terms of the potential for scientific projects to deliver justification; and explaining discomfort with science developed with artificial intelligence, given the difficulty of satisfying the discursive requirement on justification (since the reasons for results generated with artificial intelligence are often opaque). Perhaps an account of progress in philosophy could be developed on the grounds that progress for an intellectual domain need not be based on factive requirements. I leave these ideas, and would be delighted if someone were to pursue them. The heart of science will be beating, I hope healthily, long into the future.

Acknowledgments

I am very grateful to Ina Jäntgen, Benjamin Chin-Yee, Johanna Stüger, Nora Hangel, Corey Dethier, Inkeri Koskinen, Chrysostomos Mantzavinos, and Peter Vickers for reading drafts of this book at various stages of development. They provided extremely detailed and insightful comments. Three anonymous reviewers also read the full manuscript and provided many helpful comments, questions, and positive inspiration.

One of the great joys during the writing of this book was my research group of incredible PhD students, postdoctoral fellows, and visiting students. All of them contributed to this book by reading and discussing draft chapters: Adrian Erasmus, Zinhle Mncube, Hamed Tabatabaei Ghomi, Cristian Larroulet Philippi, Adrià Segarra, Oliver Holdsworth, Ina Jäntgen, Sophia Crüwell, Benjamin Chin-Yee, Florence Robinson Adams, Arthur Harris, Matthew Gummess, Claudio Davini, and Michaela Egli.

Tarun Menon has been a friend for nearly twenty years. During this time we have published three articles together, two of which I drew on extensively for this book.

Many colleagues read particular chapters and gave helpful feedback, including T. Y. Branch, Matthew Brown, Viktoria Cologna, Catherine Elgin, Mathias Frisch, Katherine Furman, Mikkel Gerken, Hannah Hilligardt, Hanna Metzen, Robert Northcott, Kristina Rolin, Jamie Shaw, Michael Strevens, and Torsten Wilholt.

I am grateful to a very broad range of friends and colleagues for discussion and correspondence about aspects of the book, including Anna Alexandrova, Douglas Allchin, Alexander Bird, Martin Carrier, Nancy Cartwright, Eugene Chua, David Colaço, Finnur Dellsén, Alkistis Elliott-Graves, Sam Fletcher, Andrew Forcehimes, Mikkel Gerken, Sara Green, Stephan Hartmann, Catherine Herfeld, Daniel Herrmann, Marie Kaiser,

Kareem Khalifa, Donal Khosrowi, Martin King, Charlie Kurth, Chiara Lisciandra, Freek Oude Maatman, Edouard Machery, Boaz Miller, Teru Miyake, Tzvetan Moev, Dario Mortini, James Norton, Thi Nguyen, Wendy Parker, Ignacio Ojea Quintana, Jesper Lundsfryd Rasmussen, Jan-Willem Romeijn, Joe Roussos, Andrew Schroeder, Miriam Solomon, Mauricio Suarez, Winnie Sung, Somogy Varga, Denis Walsh, Eric Winsberg, and Joeri Witteveen.

A fellowship at Leibniz University Hannover provided me with one year to devote to the completion of this book. I am grateful to Mathias Frisch and Torsten Wilholt for giving me this opportunity via their Socrates Centre funded by the Deutsche Forschungsgemeinschaft (DFG, German Research Foundation—Project 470816212/KFG43). I also received research support from the University of Johannesburg; Hughes Hall, Cambridge; the Department of History and Philosophy of Science, University of Cambridge; and Nanyang Technological University, Singapore.

Parts of chapter 1 first appeared as "The Difference-to-Inference Model for Values in Science" (with Tarun Menon), in *Res Philosophica* 100, no. 4 (2023): 423–47, though §1.1, §1.2, and §1.4 are entirely new; chapter 3 was published originally as "Sisyphean Science: Why Value-Freedom Is Worth Pursuing" (with Tarun Menon), in *European Journal for Philosophy of Science* 13 (2023): 48; chapter 5 was published originally as "Justifying Scientific Progress" in *Philosophy of Science* 91 (2024): 543–60; and chapter 7 will be published as "Fast Science" in *British Journal for the Philosophy of Science*.

References

Abaluck, J. et al. 2022. "Impact of Community Masking on COVID-19: A Cluster-Randomized Trial in Bangladesh." *Science* 375 (6577). eabi9069.
Adamson, Peter, and Jonardon Ganeri. 2020. *Classical Indian Philosophy*. Oxford University Press.
Aikhenvald, Alexandra. 2004. *Evidentiality*. Oxford University Press.
Alexandrova, Anna. 2018. "Can the Science of Well-Being Be Objective?" *British Journal for the Philosophy of Science* 69 (2): 421–45.
Anderson, Elizabeth. 2004. "Uses of Value Judgments in Science: A General Argument, with Lessons from a Case Study of Feminist Research on Divorce." *Hypatia* 19 (1): 1–24.
Antognazza, Maria Rosa. 2008. *Leibniz: An Intellectual Biography*. Cambridge University Press.
Appiah, Kwame Anthony. 2017. *As If: Idealization and Ideals*. Harvard University Press.
Arneson, Richard. 2005. "Sophisticated Rule Consequentialism: Some Simple Objections." *Philosophical Issues* 15:235–51.
Atkinson, Tony. 1998. *Poverty in Europe*. Blackwell.
Avery, Oswald, Colin MacLeod, and Maclyn McCarty. 1944. "Studies on the Chemical Nature of the Substance Inducing Transformation of Pneumococcal Types: Induction of Transformation by a Deoxyribonucleic Acid Fraction Isolated from Pneumococcus Type III." *Journal of Experimental Medicine* 79 (2): 137–58.
Ballarini, Cristina. 2022. "Epistemic Blame and the New Evil Demon Problem." *Philosophical Studies* 179 (8): 2475–505.
Bárdos, Dániel, and Adam Tamas Tuboly. 2025. *Science, Pseudoscience, and the Demarcation Problem*. Elements in Philosophy of Science. Cambridge University Press.
Barnes, Eric. 2008. *The Paradox of Predictivism*. Cambridge University Press.
Baumeister, R. F., E. Bratslavsky, M. Muraven, and D. M. Tice. 1998. "Ego Depletion: Is the Active Self a Limited Resource?" *Journal of Personality and Social Psychology* 74 (5): 1252–65.

Beebe, James, and Finnur Dellsén. 2020. "Scientific Realism in the Wild: An Empirical Study of Seven Sciences and History and Philosophy of Science." *Philosophy of Science* 87:336–64.
Berker, Selim. 2013. "The Rejection of Epistemic Consequentialism." *Philosophical Issues* 23 (1): 363–87.
Betz, Gregor. 2013. "In Defence of the Value Free Ideal." *European Journal for Philosophy of Science* 3:207–20.
Birch, Jonathan. 2021. "Science and Policy in Extremis: The UK's Initial Response to Covid-19." *European Journal for Philosophy of Science* 11:90.
Bird, Alexander. 2007. "What Is Scientific Progress?" *Noûs* 41:64–89.
Bird, Alexander. 2008. "Scientific Progress as Accumulation of Knowledge: A Reply to Rowbottom." *Studies in History and Philosophy of Science Part A* 39:279–81.
Bird, Alexander. 2011. "The Epistemological Function of Hill's Criteria." *Preventive Medicine* 53 (4–5): 242–45.
Bird, Alexander. 2021. "Understanding the Replication Crisis as a Base Rate Fallacy." *British Journal for the Philosophy of Science* 72 (4): 965–93.
Bird, Alexander. 2022. *Knowing Science*. Oxford University Press.
Bishop, Michael, and J. D. Trout. 2005. *Epistemology and the Psychology of Human Judgment*. Oxford University Press.
Bostrom, Nick. 2013. "Existential Risk Prevention as Global Priority." *Global Policy* 4:15–31.
Bradford, Gwen. 2015. *Achievement*. Oxford University Press.
Bradford Hill, Austin. 1965. "The Environment and Disease: Association or Causation?" *Proceedings of the Royal Society of Medicine* 58 (5): 295–300.
Bradley, Richard, Franz Dietrich, and Christian List. 2014. "Aggregating Causal Judgments." *Philosophy of Science* 81 (4): 491–515.
Bradley, Richard, and Katie Steele. 2015. "Making Climate Decisions." *Philosophy Compass* 10–11:799–810.
Bradley, Seamus, and Katie Steele. 2016. "Can Free Evidence Be Bad? Value of Information for the Imprecise Probabilist." *Philosophy of Science* 83 (1): 1–28.
Bright, Liam K. 2018. "Du Bois' Democratic Defence of the Value Free Ideal." *Synthese* 195:2227–245.
Broome, John. 2012. *Climate Matters: Ethics in a Warming World*. Norton.
Brown, Matthew. 2013. "Values in Science Beyond Underdetermination and Inductive Risk." *Philosophy of Science* 80:829–39.
Brown, Matthew. 2019. "Is Science Really Value Free and Objective? From Objectivity to Scientific Integrity." In *What Is Scientific Knowledge? An Introduction to Contemporary Epistemology of Science*, edited by Kevin McCain and Kostas Kampourakis, 226–42. New York: Routledge.
Brown, Matthew. 2020. *Science and Moral Imagination: A New Ideal for Values in Science*. University of Pittsburgh Press.

Brown, Matthew, and Joyce Havstad. 2016. "The Disconnect Problem, Scientific Authority, and Climate Policy." *Perspectives on Science* 25 (1): 67–94.
Brown, Matthew, and Jacob Stegenga. 2023. "The Validity of the Argument from Inductive Risk." *Canadian Journal of Philosophy* 53 (2): 187–90.
Bueter, Anke. 2022. "Bias as an Epistemic Notion." *Studies in History and Philosophy of Science* 91:307–15.
Callender, Craig. 2020. "Can We Quarantine the Quantum Blight?" in *Scientific Realism and the Quantum*, edited by Steven French and Juha Saatsi. Oxford University Press, 57–77.
Carrier, Martin. 2013. "Values and Objectivity in Science: Value-Ladenness, Pluralism and the Epistemic Attitude." *Science & Education* 22 (10): 2547–68.
Cartwright, Nancy. 1983. *How the Laws of Physics Lie*. Oxford University Press.
Cartwright, Nancy. 2006. "Well-Ordered Science: Evidence for Use." *Philosophy of Science* 73 (5): 981–90.
Cartwright, Nancy, Jeremy Hardie, Eleonora Montuschi, Matthew Soleiman, and Ann Thresher. 2022. *The Tangle of Science: Reliability Beyond Method, Rigour, and Objectivity*. Oxford University Press.
Chadwick, James. 1932. "Possible Existence of a Neutron." *Nature* 129:312.
Chalmers, David. 2022. *Reality +: Virtual Worlds and the Problems of Philosophy*. Norton.
Chang, Hasok. 2022. *Realism for Realistic People: A New Pragmatist Philosophy of Science*. Cambridge University Press.
Chargaff, Erwin. 1951. "Structure and Function of Nucleic Acids as Cell Constituents." *Federation Proceedings* 10:654–59.
Chrisman, Matthew. 2022. *Belief, Agency, and Knowledge: Essays on Epistemic Normativity*. Oxford University Press.
Claveau, François, and Olivier Grenier. 2019. "The Variety-of-Evidence Thesis: A Bayesian Exploration of Its Surprising Failures." *Synthese* 196:3001–28.
Clifford, William. 1877. "The Ethics of Belief." *Contemporary Review* 29:290–309.
Collins, Harry, and Robert Evans. 2017. *Why Democracies Need Science*. Polity.
Conix, Stijn. 2020. "Enzyme Classification and the Entanglement of Values and Epistemic Standards." *Studies in History and Philosophy of Science Part A* 84:37–45.
Curie, Marie. 1923. *Autobiographical Notes*. Translated by Charlotte and Vernon Kellogg. Macmillan.
Dang, Haixin, and Liam Bright. 2021. "Scientific Conclusions Need Not Be Accurate, Justified, or Believed by Their Authors." *Synthese* 199:8187–203.
Daston, Lorraine, and Peter Galison. 2007. *Objectivity*. Zone Books.
Davidson, Donald. 2000. "Truth Rehabilitated." In *Rorty and His Critics*, edited by Robert Brandom. Blackwell, 65–73.
Dawid, Richard, Stephan Hartmann, and Jan Sprenger. 2015. "The No Alternatives Argument." *British Journal for the Philosophy of Science* 66 (1): 213–34.

Delbrück, Max. 1978. Interview by Carolyn Harding. California Institute of Technology Archives.

Dellsén, Finnur. 2016. "Scientific Progress: Knowledge Versus Understanding." *Studies in History and Philosophy of Science* 56:72–83.

Dellsén, Finnur. 2018. "Scientific Progress: Four Accounts." *Philosophy Compass* 13:e12525.

Dellsén, Finnur. 2023. "Scientific Progress Without Justification." In *Scientific Understanding and Representation: Modeling in the Physical Sciences*, edited by Insa Lawler, Kareem Khalifa, and Elay Shech. Routledge.

de Melo-Martín, Inmaculada. 2024. "Just the Facts Ma'am: Some Concerns About the Identification of Future-Proof Science." *Metascience* 33: 5–10.

de Melo-Martín, Inmaculada, and Kristen Intemann. 2016. "The Risk of Using Inductive Risk to Challenge the Value-Free Ideal." *Philosophy of Science* 83 (4): 500–20.

de Melo-Martín, Inmaculada, and Kristen Intemann. 2018. *The Fight Against Doubt: How to Bridge the Gap Between Scientists and the Public*. Oxford University Press.

de Regt, Henk. 2017. *Understanding Scientific Understanding*. Oxford University Press.

Dethier, Corey. 2022. "Science, Assertion, and the Common Ground." *Synthese*, 200.

Dethier, Corey. 2023. "The Cooperative Origins of Epistemic Rationality?" *Erkenntnis* 88 (3): 1269–88.

Dethier, Corey. 2024. "Contrast Classes and Agreement in Climate Modeling." *European Journal for Philosophy of Science* 14:14.

Dogramaci, Sinan. 2015. "Communist Conventions for Deductive Reasoning." *Nous* 49 (4): 776–99.

Douglas, Heather. 2000. "Inductive Risk and Values in Science." *Philosophy of Science* 67 (4): 559–79.

Douglas, Heather. 2004. "The Irreducible Complexity of Objectivity." *Synthese* 138 (3): 453–73.

Douglas, Heather. 2009. *Science, Policy, and the Value-Free Ideal*. University of Pittsburgh Press.

Douglas, Heather. 2014. "Pure Science and the Problem of Progress." *Studies in History and Philosophy of Science* 46:55–63.

Douglas, Heather. 2016. "Values in Science." In *Oxford Handbook of Philosophy of Science*, edited by Paul Humphreys. Oxford University Press, 609–31.

Douglas, Heather, and Kevin Elliott. 2022. "Addressing the Reproducibility Crisis: A Response to Hudson." *Journal for General Philosophy of Science* 53:201–9.

Douven, Igor. 2006. "Assertion, Knowledge, and Rational Credibility." *Philosophical Review* 115:449–85.

Duhem, Pierre. (1969) 1908. *To Save the Phenomena*. Translated by E. Dolan and C. Maschler. University of Chicago Press.

Dutilh Novaes, Catarina. 2021. *The Dialogical Roots of Deduction: Historical, Cognitive, and Philosophical Perspectives on Reasoning*. Cambridge University Press.

Einstein, Albert, and Leo Szilard. 1939. "Einstein-Szilard Letter to President Roosevelt." Available online and in Roosevelt Library, Hyde Park, NY.

Elgin, Catherine. 2017. *True Enough*. MIT Press.

Ellenberg, Jordan. 2012. "Mochizuki on ABC." *Quomodocumque* (blog).

Elliott, Kevin. 2011. *Is a Little Pollution Good for You?: Incorporating Societal Values in Environmental Research*. Oxford University Press.

Elliott, Kevin. 2017. *A Tapestry of Values: An Introduction to Values in Science*. Oxford University Press.

Elliott, Kevin. 2022. "A Taxonomy of Transparency in Science." *Canadian Journal of Philosophy* 52 (3): 342–55.

Elliott, Kevin, and Daniel McKaughan. 2014. "Nonepistemic Values and the Multiple Goals of Science." *Philosophy of Science* 81:1–21.

Erasmus, Adrian. 2023. "The Bias Dynamics Model: Correcting for Meta-Biases in Therapeutic Prediction." *Philosophy of Science* 90 (5): 1204–13.

Fahrbach, Ludwig. 2011. "Theory Change and Degrees of Success." *Philosophy of Science* 78:1283–92.

Ferguson, N. M., D. Laydon, G. Nedjati-Gilani et al. (2020). "Report 9: Impact of Non-pharmaceutical Interventions (NPIs) to Reduce COVID-19 Mortality and Healthcare Demand." *Imperial College COVID Response Team. March 16*.

Fermi, Enrico. 1934. "Possible Production of Elements of Atomic Number Higher than 92." *Nature* 133:898–99.

Feyerabend, Paul. 1965. "Reply to Criticism: Comments on Smart, Sellars and Putnam." *Proceedings of the Boston Colloquium for the Philosophy of Science*, 223–61.

Feyerabend, Paul. 1975. *Against Method*. Verso.

Fine, Arthur. 1986. "Unnatural Attitudes: Realist and Instrumentalist Attachments to Science." *Mind* 95 (378): 149–79.

Fisher, Ronald. 1958. "Cancer and Smoking." *Nature* 182:596.

Fitelson, Branden. 1999. "The Plurality of Bayesian Measures of Confirmation and the Problem of Measure Sensitivity." *Philosophy of Science* 66:S362–78.

Fleisher, Will. 2021. "Endorsement and Assertion." *Noûs* 55 (2): 3363–84.

Floridi, Luciano. 2011. *The Philosophy of Information*. Oxford University Press.

Frances, Allen. (2015). "The Crisis of Confidence in Medical Research." *Huffington Post*, January 12, 2015.

Freeman, Derek. 1983. *Margaret Mead and Samoa: The Making and Unmaking of an Anthropological Myth*. Harvard University Press.

Friedman, Daniel, and Dunja Šešelja. 2023. "Scientific Disagreements, Fast Science and Higher-Order Evidence." *Philosophy of Science* 90:937–57.

Frisch, Mathias. 2015. "Predictivism and Old Evidence: A Critical Look at Climate Model Tuning." *European Journal for Philosophy of Science* 5 (2): 171–90.

Frisch, Mathias. 2020. "Uncertainties, Values, and Climate Targets." *Philosophy of Science* 87 (5): 979–90.
Frisch, Otto. 1939. "Physical Evidence for the Division of Heavy Nuclei Under Neutron Bombardment." *Nature* 143:276.
Fuller, Steve. 1988. *Social Epistemology*. Indiana University Press.
Gardiner, Stephen. 2006. "A Core Precautionary Principle." *Journal of Political Philosophy* 14:33-60.
Gerken, Mikkel. 2011. "Warrant and Action." *Synthese* 178:529–47.
Gerken, Mikkel. 2017a. "Against Knowledge-First Epistemology." In *Knowledge-First Approaches in Epistemology and Mind*, edited by J. Adam Carter, Emma C. Gordon, and Benjamin Jarvis, 46–71. Oxford University Press.
Gerken, Mikkel. 2017b. *On Folk Epistemology: How We Think and Talk About Knowledge*. Oxford University Press.
Gerken, Mikkel. 2022. *Scientific Testimony: Its Role in Science and Society*. Oxford University Press.
Gert, Bernard. 1998. *Morality: Its Nature and Justification*. Oxford University Press.
Gibbons, John. 2013. *The Norm of Belief*. Oxford University Press.
Glüer, Katrin, and Åsa Wikforss. 2009. "Against Content Normativity." *Mind* 118:31–70.
Godfrey-Smith, Peter. 2003. *Theory and Reality: An Introduction to the Philosophy of Science*. University of Chicago Press.
Goldberg, Sanford C. 2015. *Assertion: On the Philosophical Significance of Assertoric Speech*. Oxford University Press.
Goldenberg, Maya. 2021. *Vaccine Hesitancy: Public Trust, Expertise, and the War on Science*. University of Pittsburgh Press.
Goldman, Alvin. 1997. "Science, Publicity, and Consciousness." *Philosophy of Science* 64:525–45.
Good, I. J. 1967. "On the Principle of Total Evidence." *British Journal for the Philosophy of Science* 17:319–21.
Greco, Daniel. 2019. "Justifications and Excuses in Epistemology." *Noûs* 55 (3): 517-37.
Grice, Paul. 1989. *Studies in the Way of Words*. Harvard University Press.
Grünbaum, Adolf. 1979. "Is Freudian Psychoanalytic Theory Pseudo-Scientific by Karl Popper's Criterion of Demarcation?" *American Philosophical Quarterly* 16 (2): 131–41.
Habermas, Jürgen. 1996. *Between Facts and Norms: Contributions to a Discourse Theory of Law and Democracy*. Translated by W. Rehg. MIT Press.
Hacking, Ian. 1983. *Representing and Intervening*. Cambridge University Press.
Hacking, Ian. 1988. "Origins of Randomization in Experimental Design." *Isis* 79 (3): 427–51.
Hacking, Ian. 2015. "Let's Not Talk About Objectivity." In *Objectivity in Science: New Perspectives from Science and Technology Studies*, edited by F. Padovani, A. Richardson, and J. Y. Tsou, 19–33. Springer.

Hahn, Otto, and Fritz Strassmann. 1939. "Über den Nachweis und das Verhalten der bei der Bestrahlung des Urans mittels Neutronen entstehenden Erdalkalimetalle" ["On the Detection and Characteristics of the Alkaline Earth Metals Formed by Irradiation of Uranium with Neutrons]. *Die Naturwissenschaften* 27 (1): 11–15.

Hall, Stephen. 2011. "Scientists on Trial: At Fault?" *Nature* 477:264–69.

Hanel, Stephani. 2015. "Lise Meitner—Fame Without a Nobel Prize." www.lindau-nobel.org/lise-meitner-fame-without-a-nobel-prize/.

Hangel, Nora, and Jutta Schickore. 2017. "Scientists' Conceptions of Good Research Practice." *Perspectives on Science* 25 (6): 766–91.

Hansson, Sven Ove. 2021. "Science and Pseudo-Science." *Stanford Encyclopedia of Philosophy*.

Hardimon, Michael. 1994. "Role Obligations." *Journal of Philosophy* 91 (7): 333–63.

Harris, KR. 2024. "Scientific Progress and Collective Attitudes." *Episteme* 21 (1): 127–46.

Havstad, Joyce. 2022. "Sensational Science, Archaic Hominin Genetics, and Amplified Inductive Risk." *Canadian Journal of Philosophy* 52 (3): 295–320.

Havstad, Joyce, and Matthew Brown. 2017a. "Inductive Risk, Deferred Decisions, and Climate Science Advising." In *Exploring Inductive Risk: Case Studies of Values in Science*, edited by Kevin Elliott and Ted Richards, 101–26. Oxford University Press.

Havstad, Joyce, and Matthew Brown. 2017b. "Neutrality, Relevance, Prescription, and the IPCC." *Public Affairs Quarterly* 31 (4): 303–24.

Hawthorne, John, and Artūrs Logins. 2021. "Graded Epistemic Justification." *Philosophical Studies* 178:1845–58.

Heal, Jane. 1978. "Common Knowledge." *Philosophical Quarterly* 28 (111): 116–31.

Heesen, Remco. 2018. "Why the Reward Structure of Science Makes Reproducibility Problems Inevitable." *Journal of Philosophy* 115 (12): 661–74.

Heesen, Remco. 2019. "The Credit Incentive to Be a Maverick." *Studies in History and Philosophy of Science* 76:5–12.

Heesen, Remco, and Liam Kofi Bright. 2021. "Is Peer Review a Good Idea?" *British Journal for the Philosophy of Science* 72 (3): 635–63.

Hershey, Alfred, and Martha Chase. 1952. "Independent Functions of Viral Proteins and Nucleic Acid in Growth of Bacteriophage." *Journal of General Physiology* 36:39–56.

Hicks, Daniel J. 2018. "Inductive Risk and Regulatory Toxicology: A Comment on de Melo-Martín and Intemann." *Philosophy of Science* 85 (1): 164–74.

Hill, Austin Bradford. 1965. "The Environment and Disease: Association or Causation?" *Proceedings of the Royal Society of Medicine* 58 (5): 295–300.

Hitzig, Zoe, and Jacob Stegenga. 2020. "The Problem of New Evidence: P-Hacking and Pre-Analysis Plans." *Diametros* 17 (66): 10–33.

Hoefer, Carl. 2020. "Scientific Realism Without the Quantum." In *Scientific Realism and the Quantum*, edited by Steven French and Juha Saatsi, 19–34. Oxford University Press.

Holman, Bennett, and Torsten Wilholt. 2022. "The New Demarcation Problem." *Studies in History and Philosophy of Science* 91:211–20.

Horn, Laurence Robert. 1972. "On the Semantic Properties of Logical Operators in English." PhD diss., University of California Los Angeles.

Hoyningen-Huene, Paul. 2013. *Systematicity: The Nature of Science*. Oxford University Press.

Huber, Franz. 2008. "Hempel's Logic of Confirmation." *Philosophical Studies* 139:181–89.

Hughes, Jeff. 2003. *The Manhattan Project: Big Science and the Atomic Bomb*. Columbia University Press.

Immerman, Daniel. 2022. "How Common Knowledge Is Possible." *Mind* 131 (523): 937–50.

Intemann, Kristen. 2005. "Feminism, Underdetermination, and Values in Science." *Philosophy of Science* 72:1001–12.

Intemann, Kristen. 2015. "Distinguishing Between Legitimate and Illegitimate Values in Climate Modeling." *European Journal for Philosophy of Science* 5:217–32.

Jackson, Frank. 1991. "Decision-theoretic Consequentialism and the Nearest and Dearest Objection." *Ethics* 101:461–82.

Jäntgen, Ina. 2023. "How to Measure Effect Sizes for Rational Decision-Making." *Philosophy of Science* 90 (5): 1183–93.

Jardine, Nick. 2003. "Whigs and Stories: Herbert Butterfield and the Historiography of Science." *History of Science* 41 (2): 127–28.

Jeffrey, Richard C. 1956. "Valuation and Acceptance of Scientific Hypotheses." *Philosophy of Science* 23 (3): 237–46.

John, Stephen. 2015. "The Example of the IPCC Does Not Vindicate the Value Free Ideal: A Response to Gregor Betz." *European Journal for Philosophy of Science* 5:1–13.

John, Stephen. 2018. "Epistemic Trust and the Ethics of Science Communication: Against Transparency, Openness, Sincerity and Honesty." *Social Epistemology* 32 (2): 75–87.

Joliot, Frédéric, and Irène Joliot-Curie. 1934. "Artificial Production of a New Kind of Radio-Element." *Nature* 133 (3354): 201–2.

Kant, Immanuel. 1785. *Groundwork of the Metaphysics of Morals*.

Kelly, Thomas. 2014. "Evidence Can Be Permissive." In *Contemporary Debates in Epistemology*, 2nd ed., edited by Matthias Steup, John Turri, and Ernest Sosa, 298–312. Wiley-Blackwell.

Kelly, Thomas. 2022. *Bias: A Philosophical Study*. Oxford University Press.

Kelp, Christoph, and Mona Simion. 2017. "Criticism and Blame in Action and Assertion." *Journal of Philosophy* 114 (2): 76–93.

Kelp, Christoph, and Mona Simion. 2021. *Sharing Knowledge: A Functionalist Account of Assertion*. Cambridge University Press.

Kenna, Aaron, and Jacob Stegenga. 2017. "Absolute Measures of Effectiveness." In *Measurement in Medicine: Philosophical Essays on Assessment and Evaluation*, edited by Leah M. McClimans, 35–51. Rowman & Littlefield.

Khalifa, Kareem. 2017. *Understanding, Explanation, and Scientific Knowledge*. Cambridge University Press.

Kitcher, Philip. 1982. *Abusing Science: The Case Against Creationism*. MIT Press.

Kitcher, Philip. 1990. "The Division of Cognitive Labor." *Journal of Philosophy* 87 (1): 5–22.

Kitcher, Philip. 1993. *The Advancement of Science: Science Without Legend, Objectivity Without Illusions*. Oxford University Press.

Kitcher, Philip. 2001. *Science, Truth, and Democracy*. Oxford University Press.

Knowles, Jonathan. 2002. "What's Really Wrong with Laudan's Normative Naturalism." *International Studies in the Philosophy of Science* 16 (2): 171–86.

Koskinen, Inkeri. 2020. "Defending a Risk Account of Scientific Objectivity." *British Journal for the Philosophy of Science* 71 (4): 1187–1207.

Kourany, Janet. 2008. "Replacing the Ideal of Value-Free Science." In *The Challenge of the Social and the Pressure of Practice: Science and Values Revisited*, edited by Martin Carrier, Don Howard, and Janet A. Kourany, 87–111. University of Pittsburgh Press.

Kourany, Janet. 2010. *Philosophy of Science After Feminism*. Oxford University Press.

Kuhn, Thomas. 1957. *The Copernican Revolution: Planetary Astronomy in the Development of Western Thought*. Harvard University Press.

Kuhn, Thomas. 1962. *The Structure of Scientific Revolutions*. University of Chicago Press.

Kukla, Quill [Rebecca]. 2019. "Infertility, Epistemic Risk, and Disease Definitions." *Synthese* 196:4409–28.

Kuorikoski, Jaakko, and Caterina Marchionni. 2016. "Evidential Diversity and the Triangulation of Phenomena." *Philosophy of Science* 83:227–47.

Kusch, Martin. 2002. *Knowledge by Agreement*. Oxford University Press.

Kvanvig, Jonathan. 2009. "Assertion, Knowledge, and Lotteries." In *Williamson on Knowledge*, edited by P. Duncan and P. Greenough, 140–60. Oxford University Press.

Lacey, Hugh. 1999. *Is Science Value Free? Values and Scientific Understanding*. Routledge.

Lackey, Jennifer. 2007. "Norms of Assertion." *Nous* 41 (4): 594–626.

Lakatos, Imre. 1978. *The Methodology of Scientific Research Programmes (Philosophical Papers*: Volume 1). Edited by John Worrall and Gregory Currie. Cambridge University Press.

Larroulet Philippi, Cristian. 2020. "Well-Ordered Science's Basic Problem." *Philosophy of Science* 87 (2): 365–75.

Larroulet Philippi, Cristian. 2022. "There Is Cause to Randomize." *Philosophy of Science* 89 (1): 152–70.

Latour, Bruno. 2004. "Why Has Critique Run Out of Steam? From Matters of Fact to Matters of Concern." *Critical Inquiry* 30 (2): 225–48.

Laudan, Larry. 1977. *Progress and Its Problems: Towards a Theory of Scientific Growth*. University of California Press.

Laudan, Larry. 1981. "A Confutation of Convergent Realism." *Philosophy of Science* 48 (1): 19–49.

Laudan, Larry. 1983. "The Demise of the Demarcation Problem." In *Physics, Philosophy and Psychoanalysis*, edited by R. S. Cohen and L. Laudan. Boston Studies in the Philosophy of Science, vol. 76. Springer.

Laudan, Larry. 1984. *Science and Values: The Aims of Science and Their Role in Scientific Debate*. University of California Press.

Laudan, Larry. 1990. "Aim-less Epistemology?" *Studies in History and Philosophy of Science Part A* 21 (2): 315–322.

Laudan, Larry. 1996. *Beyond Positivism and Relativism: Theory, Method, and Evidence*. Westview.

Lederberg, Joshua. 1958. Nobel Lecture. www.nobel.se.

Lederberg, Joshua, and Edward Tatum. 1946. "Gene Recombination in Eschericia coli." *Nature* 58:558.

Lederman, Harvey. 2018. "Uncommon Knowledge." *Mind* 127:1069–1105.

Lehrer, Keith, and Carl Wagner. 1981. *Rational Consensus in Science and Society*. D. Reidel.

Leibniz, G. W. 1678–79. *Studia ad Felicitatem Dirigenda*.

Leibniz, G. W. (c. 1702–3) 1988. "Meditation on the Common Concept of Justice." In *Leibniz: Political Writings. Cambridge Texts in the History of Political Thought*, edited by P. Riley, 45–64. Cambridge University Press.

Levi, Isaac. 1967. *Gambling with Truth: An Essay on Induction and the Aims of Science*. Routledge & Kegan Paul.

Levy, Arnon. Forthcoming. "Thought Experiments Repositioned." In *Methods in the Philosophy of Science: A User's Guide*, edited by Adrian Currie and Sophie Veigl. MIT Press.

Lewis, David. 1969. *Convention*. Harvard University Press.

Li, Chenyang. 2023. "The Sequential Problem of the Eight Human Aims in the Great Learning." *Philosophy East and West* 73 (2): 326–44.

Lipton, Peter. 2004. *Inference to the Best Explanation*, 2nd ed. Routledge.

Lipton, Peter. 2005. "The Medawar Lecture 2004: The Truth About Science." *Philosophical Transactions of the Royal Society: Biological Sciences* 360 (1458): 1259–69.

Littlejohn, Clayton. Forthcoming. "A Plea for Epistemic Excuses." In *The New Evil Demon*, edited by F. Dorsch and J. Dutant. Oxford University Press.

Longino, Helen. 1990. *Science as Social Knowledge: Values and Objectivity in Scientific Inquiry*. Princeton University Press.

Longino, Helen. 2002. *The Fate of Knowledge*. Princeton University Press.

Longino, Helen. 2004. "How Values Can Be Good for Science." In *Science, Values, and Objectivity*, edited by Peter Machamer and Gereon Walters, 127–42. University of Pittsburgh Press.

Lynch, Michael. 2019. *Know-It-All Society: Truth and Arrogance in Political Culture*. Norton.

Madison, B. J. C. 2018. "On Justifications and Excuses." *Synthese* 195 (10): 4551–62.

Magnus, P. D., and Craig Callender. 2004. "Realist Ennui and the Base Rate Fallacy." *Philosophy of Science* 71 (3): 320–38.

Mandelkern, Matthew, and Kevin Dorst. 2022. "Assertion Is Weak." *Philosophers' Imprint* 22:19.

Mantzavinos, Chrysostomos. 2024. *The Constitution of Science*. Cambridge University Press.

Massimi, Michela. 2016. "Three Tales of Scientific Success." *Philosophy of Science* 83:757–67.

Massimi, Michela. 2022a. *Perspectival Realism*. Oxford University Press.

Massimi, Michela. 2022b. "Perspectives on Scientific Progress." *Nature Physics* 18:604–6.

Mead, Margaret. 1928. *Coming of Age in Samoa: A Psychological Study of Primitive Youth for Western Civilisation*. William Morrow.

Meitner, Lise, and Otto Frisch. 1939. "Disintegration of Uranium by Neutrons: A New Type of Nuclear Reaction." *Nature* 143 (3615): 239.

Menon, Tarun, and Jacob Stegenga. 2023. "Sisyphean Science: Why Value-Freedom Is Worth Pursuing." *European Journal for Philosophy of Science* 13:48.

Merton, Robert. 1957. "Priorities in Scientific Discovery." *American Sociological Review* 22:635–59.

Merton, Robert. 1973. *The Sociology of Science: Theoretical and Empirical Investigations*. University of Chicago Press.

Merton, Robert K. (1942) 1973. "The Normative Structure of Science." In *The Sociology of Science: Theoretical and Empirical Investigations*. University of Chicago Press, 267–78.

Mill, John Stuart. 1859. *On Liberty*. John W. Parker and Son.

Miller, Boaz. 2013. "When Is Consensus Knowledge Based? Distinguishing Shared Knowledge from Mere Agreement." *Synthese* 190:1293–1316.

Miller, Boaz. 2016. "Scientific Consensus and Expert Testimony in Courts: Lessons from the Bendectin Litigation." *Foundations of Science* 21:15–33.

Miller, Boaz. 2021. "When Is Scientific Dissent Epistemically Inappropriate?" *Philosophy of Science* 88 (5): 918–28.

Mills, Charles. 2005. "'Ideal Theory' as Ideology." *Hypatia* 20 (3): 165–84.

Mitchell, Sandra. 2020. "Through the Fractured Looking Glass." *Philosophy of Science* 87 (5): 771–92.

Mizrahi, Moti. 2016. "The History of Science as a Graveyard of Theories: A Philosophers' Myth?" *International Studies in the Philosophy of Science* 30 (3): 263–78.

Mizrahi, Moti, and Wesley Buckwalter. 2014. "The Role of Justification in the Ordinary Concept of Scientific Progress." *Journal for General Philosophy of Science* 45 (1): 151–66.

Nagel, Thomas. 1986. *The View from Nowhere*. Oxford University Press.

Nguyen, Thi. 2023. "Hostile Epistemology." *Social Philosophy Today* 39:9–32.

Nietzsche, Friedrich. 1968. *The Will to Power*. Vintage Books.

Niiniluoto, Ilkka. 2014. "Scientific Progress as Increasing Verisimilitude." *Studies in History and Philosophy of Science Part A* 46:73–77.

Noddack, Ida. 1934. "Über das Element 93." *Zeitschrift für Angewandte Chemie*. 47 (37): 653.

Northcott, Robert. 2022. "Pandemic Modeling, Good and Bad." *Philosophy of Medicine* 3 (1).

Norton, John. 2000. "How We Know About Electrons." In *After Popper, Kuhn and Feyerabend*, edited by R. Nola and H. Sankey, 67–97. Kluwer.

Norton, John. 2021. *The Material Theory of Induction*. University of Calgary Press.

Okasha, Samir. 2011. "Theory Choice and Social Choice: Kuhn Versus Arrow." *Mind* 477:83–115.

Oreskes, Naomi. 2004. "The Scientific Consensus on Climate Change." *Science* 306:1686.

Oreskes, Naomi. 2019. *Why Trust Science?* Princeton University Press.

Oreskes, Naomi, and Erik Conway. 2010. *Merchants of Doubt*. Bloomsbury.

Pamuk, Zeynep. 2021. *Politics and Expertise: How to Use Science in a Democratic Society*. Princeton University Press.

Papineau, David. 2021. "The Disvalue of Knowledge." *Synthese* 198:5311-5332.

Parker, Wendy. 2014. "Values and Uncertainties in Climate Prediction, Revisited." *Studies in History and Philosophy of Science Part A* 46:24–30.

Peters, Uwe. 2020. "Values in Science: Assessing the Case for Mixed Claims." *Inquiry* 66 (6): 965–76.

Peters, Uwe. 2021. "Illegitimate Values, Confirmation Bias, and Mandevillian Cognition in Science." *British Journal for the Philosophy of Science* 72 (4): 1061–81.

Pielke, Roger. 2007. *The Honest Broker: Making Sense of Science in Policy and Politics*. Cambridge University Press.

Polanyi, Michael. 1962. "The Republic of Science: Its Political and Economic Theory." *Minerva* 1:54-74.

Popper, Karl. 1959. *The Logic of Scientific Discovery*. Hutchinson.

Popper, Karl. 1963. *Conjectures and Refutations*. Routledge.

Popper, Karl. 1994. *The Myth of the Framework*. Routledge.

Posner, Richard. 2004. *Catastrophe: Risk and Response*. Oxford University Press.

Potochnik, Angela. 2017. *Idealization and the Aims of Science*. University of Chicago Press.

Prendini, Lorenzo. 2021. "Taxa Dedicated to Norman I. Platnick." *Entomologica Americana* 126 (1–4): 101–7.

Price, Huw. 2003. "Truth as Convenient Friction." *Journal of Philosophy* 100 (4): 167–90.
Proctor, Robert N. 1991. *Value Free Science: Purity and Power in Modern Knowledge*. Harvard University Press.
Putnam, Hilary. (1974) 1991. "The 'Corroboration' of Theories." In Schilpp (1974), *The Philosophy of Karl Popper*, vol.1, edited by Paul Schilpp, 221–40. Open Court. Republished with Retrospective Note in *The Philosophy of Science*, edited by Richard Boyd, Philip Gasper, and J. D. Trout, 121–38. MIT Press.
Quong, Jonathan. 2011. *Liberalism Without Perfection*. Oxford University Press.
Rawls, John. 1971. *A Theory of Justice*. Harvard University Press.
Reid, Robert William. 1974. *Marie Curie*. New American Library.
Resnik, David, and Kevin Elliott. 2023. "Science, Values, and the New Demarcation Problem." *Journal for General Philosophy of Science* 54:259–86.
Rhodes, Richard. 1986. *The Making of the Atomic Bomb*. Simon & Schuster. References are to the 1988 edition.
Richardson, Alan. 2023. *Logical Empiricism as Scientific Philosophy*. Cambridge University Press.
Rolin, Kristina. 2017. "Scientific Community: A Moral Dimension." *Social Epistemology* 31 (5): 468–83.
Rolin, Kristina. 2020. "Group Disagreement in Science." In *The Epistemology of Group Disagreement*, edited by Fernando Broncano-Berrocal and J. Adam Carter, 163–83. Routledge.
Romero, Felipe. 2017. "Novelty Versus Replicability: Virtues and Vices in the Reward System of Science." *Philosophy of Science* 84 (5): 1031–43.
Rorty, Richard. 1995. "Is Truth a Goal of Enquiry? Davidson Vs. Wright." *Philosophical Quarterly* 45:281–300.
Rorty, Richard. 1998. *Truth and Progress: Philosophical Papers, Volume 3*. Cambridge University Press.
Rosenkranz, Sven. 2023. "Problems for Factive Accounts of Assertion." *Noûs* 57:128–43.
Ross, Lewis. 2020. "How Intellectual Communities Progress." *Episteme*, 1–19.
Rowbottom, Darrell. 2008. "N-rays and the Semantic View of Scientific Progress." *Studies in History and Philosophy of Science* 39:277–78.
Rowbottom, Darrell. 2023. *Scientific Progress*. Cambridge University Press.
Rubin, Hannah, and Mike Schneider. 2021. "Priority and Privilege in Scientific Discovery." *Studies in History and Philosophy of Science* 89:202–11.
Rudner, Richard. 1953. "The Scientist Qua Scientist Makes Value Judgements." *Philosophy of Science* 20 (1): 1–6.
Rumfitt, Ian. 2003. "Savoir Faire." *Journal of Philosophy* 100 (3): 158–66.
Ruphy, Stephanie. 2006. "Empiricism All the Way Down: A Defense of the Value-Neutrality of Science in Response to Helen Longino's Contextual Empiricism." *Perspectives on Science* 14 (2): 189–214.

Ruphy, Stephanie. 2016. *Scientific Pluralism Reconsidered: A New Approach to the (Dis)Unity of Science*. University of Pittsburgh Press.
Russell, Denise. 1983. "Anything Goes." *Social Studies of Science* 13 (3): 437–64.
Sandin, Per. 2009. "Supreme Emergencies Without the Bad Guys." *Philosophia* 37:153–67.
Schechter, Joshua. 2017. "No Need for Excuses: Against Knowledge-First Epistemology and the Knowledge Norm of Assertion." In *Knowledge-First Approaches in Epistemology and Mind*, edited by J. Adam Carter, Emma C. Gordon, and Benjamin Jarvis. Oxford University Press.
Schindler, Samuel. 2015. "Scientific Discovery: That-Whats and What-Thats." *Ergo* 2 (6): 123–48.
Schroeder, Andrew. 2017. "Using Democratic Values in Science: An Objection and (Partial) Response." *Philosophy of Science* 84 (5): 1044–54.
Schroeder, Andrew. 2021. "Democratic Values: A Better Foundation for Public Trust in Science." *British Journal for the Philosophy of Science* 72 (2): 545–62.
Schroeder, Andrew. 2022. "An Ethical Framework for Presenting Scientific Results to Policy-Makers." *Kennedy Institute of Ethics Journal* 32 (1): 33–67.
Schupbach, Jonah. 2018. "Robustness Analysis as Explanatory Reasoning." *British Journal for the Philosophy of Science* 69 (1): 275–300.
Sen, Amartya. 2009. *The Idea of Justice*. Belknap.
Shan, Yafeng. 2019. "A New Functional Approach to Scientific Progress." *Philosophy of Science* 86:739–58.
Shankman, Paul. 2009. *The Trashing of Margaret Mead: Anatomy of an Anthropological Controversy*. University of Wisconsin Press.
Shapin, Steven, and Simon Schaffer. 1985. *Leviathan and the Air-Pump*. Princeton University Press.
Shaw, Jamie. 2020. "Feyerabend and Manufactured Disagreement: Reflections on Expertise, Consensus, and Science Policy." *Synthese* 198 (25): 6053–84.
Shaw, Jamie. 2022a. "On the Very Idea of Pursuitworthiness." *Studies in History and Philosophy of Science* 91:103–12.
Shaw, Jamie. 2022b. "Revisiting the Basic/Applied Science Distinction: The Significance of Urgent Science for Science Funding Policy." *Journal for General Philosophy of Science* 53:477–99.
Sheldrick, Kyle, Gideon Meyerowitz-Katz, and Greg Tucker-Kellogg. 2022. "Plausibility of Claimed Covid-19 Vaccine Efficacies by Age: A Simulation Study." *American Journal of Therapeutics* 29 (5): e495–99.
Shogenji, Tomoji. 2012. "The Degree of Epistemic Justification and the Conjunction Fallacy." *Synthese* 184 (1): 29–48.
Siegel, Harvey. 1990. "Laudan's Normative Naturalism." *Studies in History and Philosophy of Science* 21:295–313.
Sober, Elliott. 2015. *Ockham's Razors: A User's Manual*. Cambridge University Press.

Solomon, Miriam. 2001. *Social Empiricism*. MIT Press.
Sprenger, Jan, and Stephan Hartmann. 2019. *Bayesian Philosophy of Science*. Oxford University Press.
Sprenger, Jan, and Jacob Stegenga. 2017. "Three Arguments for Absolute Outcome Measures." *Philosophy of Science* 84:840–52.
Staley, Kent, and Aaron Cobb. 2011. "Internalist and Externalist Aspects of Justification in Scientific Inquiry." *Synthese* 182 (3): 475–92.
Stanford, P. Kyle. 2015. "Catastrophism, Uniformitarianism, and a Scientific Realism Debate That Makes a Difference." *Philosophy of Science* 82 (5): 867–78.
Stanley, Jason, and Tim Williamson. 2001. "Knowing How." *Journal of Philosophy* 98 (8): 411–44.
Steel, Daniel. 2015. *Philosophy and the Precautionary Principle: Science, Evidence, and Environmental Policy*. Cambridge University Press.
Steel, Daniel. 2016a. "Accepting an Epistemically Inferior Alternative? A Comment on Elliott and McKaughan." *Philosophy of Science* 83 (4): 606–12.
Steel, Daniel. 2016b. "Climate Change and Second-Order Uncertainty: Defending a Generalized, Normative, and Structural Argument from Inductive Risk." *Perspectives on Science* 24 (6): 696–721.
Steele, Katie. 2012. "The Scientist Qua Policy Advisor Makes Value Judgments." *Philosophy of Science* 79 (5): 893–904.
Stegenga, Jacob. 2011. "The Chemical Characterization of the Gene: Vicissitudes of Evidential Assessment." *History and Philosophy of the Life Sciences* 33:105–27.
Stegenga, Jacob. 2013. "Evidence in Biology and the Conditions of Success." *Biology and Philosophy* 28 (6): 981–1004.
Stegenga, Jacob. 2015. "Theory Choice and Social Choice: Okasha Versus Sen." *Mind* 493:263–77.
Stegenga, Jacob. 2017. "Drug Regulation and the Inductive Risk Calculus." In *Exploring Inductive Risk*, edited by Kevin Elliott and Ted Richards. Oxford University Press, 17–36.
Stegenga, Jacob. 2018. *Medical Nihilism*. Oxford University Press.
Stegenga, Jacob. 2022a. "Red Herrings About Relative Measures: A Response to Hoefer and Krauss." *Studies in History and Philosophy of Science* 92:56–59.
Stegenga, Jacob. 2022b. "Sex Differences in Sexual Desire." *Philosophy of Science* 89 (5): 1094–1103.
Stegenga, Jacob. 2024. "Justifying Scientific Progress." *Philosophy of Science* 91:543–60.
Stegenga, Jacob. Forthcoming. "Fast Science." *British Journal for the Philosophy of Science*.
Stegenga, Jacob, and Tarun Menon. 2023. "The Difference-to-Inference Model of Values in Science." *Res Philosophica* 100 (4): 423–47.
Steglich-Petersen, Asbjørn. 2010. "The Truth Norm and Guidance: A Reply to Glüer and Wikforss." *Mind* 119 (475): 749–55.

Stengers, Isabelle. (2013) 2018. *Another Science Is Possible: A Manifesto for Slow Science*. Polity.

Stolley, Paul. 1991. "When Genius Errs: R. A. Fisher and the Lung Cancer Controversy." *American Journal of Epidemiology* 133 (5): 416–25.

Strevens, Michael. 2003. "The Role of the Priority Rule in Science." *Journal of Philosophy* 100 (2): 55–79.

Strevens, Michael. 2006. "The Role of the Matthew Effect in Science." *Studies in History and Philosophy of Science* 37 (2): 159–70.

Strevens, Michael. 2011. "Economic Approaches to Understanding Scientific Norms." *Episteme* 8 (2): 184–200.

Sunstein, Cass. 2005. "Cost-Benefit Analysis and the Environment." *Ethics* 115 (2): 351–85.

Sylvan, Kurk. 2020. "An Epistemic Nonconsequentialism." *Philosophical Review* 129 (1): 1–51.

Tabatabaei Ghomi, Hamed, and Jacob Stegenga. 2022. "Conventional Choices in Outcome Measures Influence Meta-Analytic Results." *Philosophy of Science* 89 (5): 949–59.

Tabatabaei Ghomi, Hamed, and Jacob Stegenga. 2023. "Simulation of Trial Data to Test Speculative Hypotheses About Medicine." In *Advances in Experimental Philosophy of Medicine*, edited by Kristien Hens and Andreas De Block, 111–28. Bloomsbury.

Thalos, Mariam. 2015. "Review of Systematicity: The Nature of Science by Paul Hoyningen-Huene." *Mind* 124:351–57.

Trouwborst, Arie. 2006. *Precautionary Rights and Duties of States*. Brill.

Toulmin, Stephen, and June Goodfield. 1961. *The Fabric of the Heavens: The Development of Astronomy and Dynamics*. University of Chicago Press.

Tucker, Aviezer. 2003. "The Epistemic Significance of Consensus." *Inquiry* 46 (4): 501–521.

Valentini, Laura. 2012. "Ideal vs. Non-Ideal Theory: A Conceptual Map." *Philosophy Compass* 7–9:654–64.

van Basshuysen, Philippe, and Lucie White. 2021. "Were Lockdowns Justified? A Return to the Facts and Evidence." *Kennedy Institute of Ethics Journal* 31 (4): 405–28.

van Basshuysen, Philippe, Lucie White, Donal Khosrowi, and Mathias Frisch. 2021. "Three Ways in Which Pandemic Models May Perform a Pandemic." *European Journal for Philosophy and Economics* 14 (1): 110–27.

van Duym, Leah Suzanne. 2014. "Informativeness, Category Membership, and the Distribution of Adjectival Past Participles." PhD dissertation, University of Minnesota.

van Fraassen, Bas. 1980. *The Scientific Image*. Oxford University Press.

Vickers, Peter. 2012. "Historical Magic in Old Quantum Theory?" *European Journal for Philosophy of Science* 2 (1): 1–19.

Vickers, Peter. 2020a. "Resisting Scientific Anti-Realism." *Metascience* 29:11–16.
Vickers, Peter. 2020b. "Disarming the Ultimate Historical Challenge to Scientific Realism." *British Journal for the Philosophy of Science* 71 (3): 987–1012.
Vickers, Peter. 2023. *Identifying Future-Proof Science*. Oxford University Press.
Vickers, Peter et al. 2024. "Development of a Novel Methodology for Ascertaining Scientific Opinion and Extent of Agreement." *PLoS ONE* 19 (12): e0313541.
Vohs, K., B. Schmeichel, S. Lohmann et al. 2020. "A Multi-site Preregistered Paradigmatic Test of the Ego Depletion Effect." *Psychological Science* 32 (10): 1566–81.
Walzer, Michael. 1973. "Political Action: The Problem of Dirty Hands." *Philosophy and Public Affairs* 2 (2): 160–80.
Walzer, Michael. 2000. *Just and Unjust Wars: A Moral Argument with Historical Illustrations*. 3rd ed. Basic Books.
Ward, Zina. 2021. "On Value-Laden Science." *Studies in History and Philosophy of Science Part A* 85:54–62.
Weiner, Matthew. 2005. "Must We Know What We Say?" *Philosophical Review* 114 (2): 227–51.
Westman, Robert. 2011. *The Copernican Question: Prognostication, Skepticism, and Celestial Order*. University of California Press.
White, Lucie, Philippe van Basshuysen, and Mathias Frisch. 2022. "When Is Lockdown Justified?" *Philosophy of Medicine* 3 (1): 1–22.
White, Roger. 2005. "Epistemic Permissiveness." *Philosophical Perspectives* 19:445–59.
Wilholt, Torsten. 2013. "Epistemic Trust in Science." *British Journal for the Philosophy of Science* 64:233–53.
Williamson, Timothy. 1996. "Knowing and Asserting." *Philosophical Review* 105 (4): 489–523.
Williamson, Timothy. 2000. *Knowledge and Its Limits*. Oxford University Press.
Winsberg, Eric. 2012. "Values and Uncertainties in the Predictions of Global Climate Models." *Kennedy Institute of Ethics Journal* 22 (2): 111–37.
Winsberg, Eric. 2018. *Philosophy and Climate Science*. Cambridge University Press.
Winsberg, Eric. 2022. "Who Is Responsible for Global Health Inequalities After Covid-19?" *Global Epidemiology*, 100081.
Winsberg, Eric, Jason Brennan, and Chris Surprenant. 2020. "How Government Leaders Violated Their Epistemic Duties During the SARS-CoV-2 Crisis." *Kennedy Institute of Ethics Journal* 30 (3–4): 215–42.
Winsberg, Eric, Jason Brennan, and Chris Surprenant. 2021. "This Paper Attacks a Strawman but the Strawman Wins: A Reply to van Basshuysen and White." *Kennedy Institute of Ethics Journal* 31 (4): 429–46.
Woodward, James. 2003. *Making Things Happen*. Oxford University Press.
Worrall, John. 2002. "*What* Evidence in Evidence-Based Medicine?" *Philosophy of Science* 69 (S3): S316–30.

Wray, K. Brad. 2018. *Resisting Scientific Realism*. Cambridge University Press.

Zollman, Kevin. 2007. "The Communication Structure of Epistemic Communities." *Philosophy of Science* 74 (5) 574–87.

Zollman, Kevin. 2018. "The Credit Economy and the Economic Rationality of Science." *Journal of Philosophy* 115 (1): 5–33.

Index

abc conjecture, 164–65
Aristotle, 19, 43, 55, 160, 166
artificial intelligence, 206, 224–25, 229
atoms, 25; atomic bombs, 15, 199–200; atomic physics, 65–66, 83–85, 128, 174–75, 177, 190, 200, 212; dense nucleus of, 8, 52, 220, 224
Avery, Oswald, 171–73, 175–76, 178–79, 181–82, 223

Bangladesh mask study, 84, 142–43, 196
Bayesianism, 20–21, 143, 154
Becquerel, Henri, 173–74
Bethe, Hans, 15
biases, 24, 29, 48, 61, 79–82, 99, 157; cognitive, 66; methodological, 15, 109, 121; unconscious, 47, 75; values, 91–92, 95, 97, 99, 110, 114–15
bifurcation points, 90–94, 99–100, 103–7, 111–15
Big Bang, 219
black holes, 176
Bohr, Niels, 113, 128, 177, 200
Boyle, Robert, 167
Bragg, William, 136
Brownian motion, 149, 169
Buddhism, 14, 87, 132, 162–63

caloric, 209
Chadwick, James, 5, 83, 223
Chase, Martha, 172–73, 175–76
climate change, 32, 40, 53, 63, 94, 102, 122, 124, 126, 189, 206
common knowledge, definition of, 28

confirmation, 4, 21, 68–69, 84, 211; formal measures of, 154–55; scientific credit and, 168, 175, 179–82; scientific progress and, 149–51, 153–59, 161, 163–64
Confucius, 22–23, 81
consensus, 28–42, 74; disagreement and, 90, 92, 225; future-proof science and, 213–14, 217–18; objections to, 55–56, 221–22; strong, 23, 28, 33–38, 40, 47, 54, 135, 169–70, 220; weak, 26, 34–36, 41–42
consequentialism, 1, 3–7, 9–13, 19, 41, 44–45, 56, 135, 210; biases and, 80; credit and, 180; value-free ideal and, 112
Copernicus, Nicolaus, 62, 131, 159–60, 211, 221
coulombs, 212–13, 222
COVID-19 pandemic, 63, 78, 94, 117, 142–44; supreme emergency, 25, 183, 188, 190–91, 196, 201–3, 205–6, 208
Crick, Francis, 5, 131, 159, 178–79, 187, 223
Curie, Marie, 13, 69, 173–74, 223

Darwin, Charles, 69, 92, 167, 170
demarcation, 3, 8, 23–24, 62–73, 76, 85–86, 161
dirty hands, 184, 192
dissent, 32, 36; importance of, 55–56
DNA: double-helical structure of, 5, 25, 30, 37, 52, 131, 159, 178–79, 187, 212, 214, 216–17, 223–25; genes and, 14–15, 25, 35, 128, 171–72, 179
Duhem, Pierre, 132

Eddington, Arthur, 153
effect sizes, 143–44, 197

ego depletion, 153–54
Einstein, Albert, 46, 68, 117, 149, 153–54, 199
electrons, 177; charge of, 212–13, 222; discovery of, 5
epicycles, 160–61
epidemiology, 39, 225; epidemiological models, 8, 63, 65, 94, 142, 144, 183, 190, 194, 202
epistemic agency, 11, 23, 42–46, 49–51, 60–61, 90
epistemic responsibility, 11, 23, 43, 45–46, 49–51, 55, 61, 90, 118, 169
epistemic values, 89–90
ether, 209
evidentiality, 141–42
evil demon scenarios, 13, 215–16
evolution, 4, 66, 68, 92
existential risks, 205–6
explanations, 147, 149, 178, 181, 229

falsificationism, 4, 8, 23, 47, 66, 72–73, 85, 129, 183, 211; criticisms of, 48, 64, 68–70. *See also* Popper, Karl
feminism: in philosophy of science, 97; in science, 114
Fermat's Last Theorem, 35–36, 164
Fermi, Enrico, 83, 174–75
Feyerabend, Paul, 10, 55–57, 138
Fisher, Ronald, 16, 38–39, 82, 109
fission, 2, 59, 174–75, 178, 212
Fleming, Alexander, 150, 177
four-color theorem, 44–45
framing problem, 92–93, 99, 101
Franklin, Rosalind, 178–79
fraud, 58, 66, 83
Freud, Sigmund, 68, 72

Galilei, Galileo, 35, 50, 55, 166, 169, 173; Leaning Tower of Pisa and, 139
gamma-ray bursts, 185, 206
genes, 14–15, 25, 35, 128, 171–73, 175–76, 178–79, 181–82, 223. *See also* DNA
genetics, 14–15, 39, 46–47, 69, 171–72; in the Soviet Union, 91
geology, 91–92, 226
germ theory, 66, 225
gold foil experiments, 84, 220
Goodall, Jane, 5

Habermas, Jürgen, 33
Hawking, Stephen, 5

hedging, 53–55, 100–103, 105–6, 109, 113, 122–23, 129
Hershey–Chase experiment, 172–73, 175–76, 178–79, 181
Higgs boson, 152–53, 169, 181
Hill's criteria, 39
Holodomor famine, 66. *See also* Russia
Hooke, Robert, 167, 169
hostile epistemology, 83–85
Hubble constant, 35, 45–46, 65, 80
Hume, David, 68

ideal theory, 57–58
idealization, 11, 177
inductive risk, 3, 93–94, 126, 188–89, 199
informativeness, 24, 39, 53–55, 103, 106, 129; definitions of, 124, 126; norm of assertion, 116, 118–20, 122–27, 142–45
Intergovernmental Panel on Climate Change, 40, 124

Joliot-Curie, Irène, 174

Kant, Immanuel, 9, 44, 66, 196–97, 223
Kepler, Johannes, 221
knowledge-possible conditions, 210, 213, 216–20, 224
Kuhn, Thomas, 4, 60, 146–47, 160, 175, 211, 214

Lakatos, Imre, 10, 69–70, 73, 161
Lamarckism, 91
L'Aquila earthquake, 15–16
Large Hadron Collider, 152–53, 181
Laudan, Larry, 12, 14, 50–51, 58–59, 62, 64–65, 133, 147, 157, 159, 161, 211, 214
Leibniz, Gottfried, 5, 14, 32, 43, 54, 169, 225
Leibniz procedure, 43–54, 57–58, 61, 65, 75–76
linguistic evidence, 21, 138–42, 188
Litvinenko, Alexander, 174
lockdowns, 122, 144, 183, 196, 201, 205
logical positivism, 67–68
Lysenko, Trofim, 91

Manhattan Project, 15, 187–88, 190, 199–200
Marx, Karl, 66, 68
masks, 8, 63; Bangladesh mask study, 84, 142–44, 196
matrix world, 215–16

INDEX   253

Mead, Margaret, 156
meaning of life, 226
Medical Nihilism, 66, 78
Meitner, Lise, 2, 59, 174–75, 212
Mendel, Gregor, 91
Mercury, precession of perihelion of, 154
Merton, Robert, 8, 66, 70, 83, 163, 168–70, 178, 183, 211, 217
metrology, 212–13
Mill, John Stuart, 55
Mochizuki, Shinichi, 164–65

naturalism, 21, 50–51
neutrons, 5, 83, 174, 223
never-ending story, 210, 213, 220–22
Newton, Isaac, 5, 169, 221
Nietzsche, Friedrich, 2
nirvana norms, 14, 59, 132–33, 161–62
Nobel Prizes, 83, 131, 159, 169, 172–76, 178–79, 215
Noble Eightfold Path, 14, 132–33, 162
Noddack, Ida, 174–75
number theory, 164–65

objectivity, 7–8, 17–18, 30, 66, 74–76, 79–80, 88, 164, 187
observability, 13, 209, 213, 216–19
Oppenheimer, Robert, 200

peer review, 8, 47, 76, 121, 183–84, 187–88, 199–201
penicillin, 150, 177
pessimistic meta-induction, 58, 133, 149, 159, 161, 209, 211, 213–15
p-hacking, 58, 83–85
phlogiston, 209
Plato, 223–24, 226
Polanyi, Michael, 46
polonium, 13, 69, 174, 176, 223
Popper, Karl, 4, 8, 10, 23, 48, 54–55, 64, 66–69, 72–73, 85, 128–29, 183, 210–11. *See also* falsificationism
poverty, 92–93, 101–2
pragmatism, 11, 133, 136, 162
precautionary principles, 204–5
predictivism, 69
preregistration, 58, 63, 67, 82–83, 85, 110
primatology, 91, 114
prime numbers, 19, 40–43, 80

provisionality thesis, 4, 25, 208–16, 219–23, 226–29
pseudoscience, 3, 23, 60, 64, 68, 73, 216
psychoanalysis, 64, 68, 72
Ptolemy: astronomical model by, 69, 177–78, 209, 211; Ptolemaic challenge, 131–32, 160–62
publication bias, 29, 109–10
pursuit-worthiness, 229

quantum mechanics, 46, 177, 218, 220; Standard Model, 152–53, 181

radiation, 173–74, 223
radium, 69, 174, 176, 223
randomized trials, 8, 16–17, 48–49, 56, 63, 71–72, 80, 82, 99, 104, 109, 129, 155, 187, 192, 195–97. *See also* Bangladesh mask study
Rawls, John, 43, 57, 112, 204
Reagan, Ronald, 77
realism-antirealism debate, 12–13, 133, 146, 149, 159, 209–10, 213, 215–18. *See also* pessimistic meta-induction
replication crisis, 67, 84, 153, 157–58, 208
Report 9, 142, 144, 201–2
retrospective benediction, 9, 14, 81, 130–31, 158–59, 161–62
revolutions, 4, 60, 131, 146, 211
Rorty, Richard, 58–59, 118, 133, 136
Royal Society, 77, 169
Russia, 6, 66, 77, 173–74, 203. *See also* Ukraine
Rutherford, Ernest, 84, 177, 220

Scientific Advisory Group for Emergencies, 202
scientific method, 51–52
seismology, 15–16
simplicity, 48, 89, 153
slow science, 188
solar system, models of, 14, 37, 55–56, 131–32, 160, 163, 177–78, 211, 214, 221, 224, 226. *See also* Copernicus, Nicolaus; Ptolemy
stratigraphy, 226
string theory, 65, 117
systematicity, 70
Szilard, Leo, 199–200

Teller, Edward, 15, 200
telomeres, 176
Thomson, J. J., 177
thought experiments, 9, 18, 21, 134, 139, 166
threat tripod, 184–93, 195–96, 198, 201
tortoise and hare, 183, 185, 187, 199–203, 206–7
toxicology, 114
transmission electron microscopy, 217
trust in science, 4, 65, 67, 74, 76–79, 92, 97, 146, 208–9, 212, 226

Ukraine, 6, 66, 140–41
understanding, 33, 43, 50, 60, 128, 147, 177
uniqueness thesis, 154–55
universe, expansion of, 35, 45–46, 65, 80

uranium, 13, 69, 173–75, 178, 200, 212
utopian aims, 14, 58–59, 133, 161

vaccines, 63, 94, 117, 202–3, 226
verifiability, 68
veritism, 19
viruses, 8, 35, 63, 135, 143, 172, 190. *See also* COVID-19 pandemic

Walzer, Michael, 184, 192
Watson, James, 5, 131, 159, 178–79, 187, 223
well-ordered science, 51, 112
Wiles, Andrew, 36

X-rays, 136

zombie unicorns, 185

www.ingramcontent.com/pod-product-compliance
Lightning Source LLC
Chambersburg PA
CBHW022047290426
44109CB00014B/1008